EARLY FORERUNNERS OF MAN

AMS PRESS
NEW YORK

EARLY FORERUNNERS OF MAN

A MORPHOLOGICAL STUDY OF THE EVOLUTIONARY ORIGIN OF THE PRIMATES

BY

W. E. Le GROS CLARK
D.Sc.(Lond.), F.R.C.S.(Eng.)

PROFESSOR ELECT OF ANATOMY, UNIVERSITY OF OXFORD;
PROFESSOR OF ANATOMY IN THE UNIVERSITY OF LONDON ; HUNTERIAN PROFESSOR AND
ARRIS AND GALE LECTURER OF THE ROYAL COLLEGE OF SURGEONS ; VICE-PRESIDENT
OF THE ANATOMICAL SOCIETY OF GREAT BRITAIN AND IRELAND ; EXAMINER
IN ANATOMY FOR THE ROYAL COLLEGE OF SURGEONS OF ENGLAND AND
FOR THE UNIVERSITY OF LONDON ; FORMERLY EXAMINER IN
ANATOMY FOR THE NATURAL SCIENCES TRIPOS OF
THE UNIVERSITY OF CAMBRIDGE

BALTIMORE

WILLIAM WOOD AND COMPANY

1934

Library of Congress Cataloging in Publication Data

Clark, Wilfrid Edward Le Gros, Sir, 1895-1971.
  Early forerunners of man.

  Reprint of the 1934 ed. published by W. Wood,
Baltimore.
  Includes bibliographies and index.
  1. Primates—Evolution. 2. Man—Origin.
3. Evolution. 4. Anatomy, Comparative. 5. Pri-
mates, Fossil. I. Title.
QL737.P9C5   1979   599'.8'0438   76-44705
ISBN 0-404-15915-X

First AMS edition published in 1979.

Reprinted from the edition of 1934, Baltimore, from an
original in the collections of the University of Chicago
Library. [Trim size of the original has been slightly altered
in this edition. Original trim size: 16 × 24.7 cm. Text area of
the original has been maintained.]

MANUFACTURED
IN THE UNITED STATES OF AMERICA

TO

GRAFTON ELLIOT SMITH, M.A., M.D., Litt.D.,
D.Sc., F.R.S.

WHO, BY HIS OWN MONUMENTAL RESEARCHES AND BY
HIS DIRECTION AND ENCOURAGEMENT OF THE RESEARCHES OF OTHERS,
HAS CONTRIBUTED SO MUCH TO OUR KNOWLEDGE OF THE
ANATOMY AND EVOLUTION OF THE PRIMATES,

THIS BOOK IS INSCRIBED

PREFACE

MAN is classed by zoologists as a member of the Primates, the latter representing one of the several Orders of Mammals. This is more than an arbitrary grouping contrived for the descriptive convenience of a catalogue. It implies that Man and all the other creatures which, in virtue of their structural organization, come into the category of the Primates had a community of origin at some distant geological epoch in the same ancestral group of mammals. There is abundant evidence to show that this is actually the case.

In the course of time, this ancestral group gave rise to diverse types endowed with evolutionary potentialities which were broadly similar but differed in significant details. Thus, while they all manifested a general tendency towards the development of large brains, opposable thumbs, flattened nails and other features characteristic of the Primates as a whole, many of them became specialized in rather divergent ways in adaptation to their particular environments. Man himself represents the terminal phase of a main line of evolution in which the progressive development of the brain has been the most distinctive feature, and, in direct correlation with this fact, he has avoided many of the aberrant specializations of other Primates and has preserved a bodily structure which is still remarkably generalized.

In order satisfactorily to study the specific problem of Man's phylogenetic origin, it is essential to visualize in proper perspective the evolutionary development of the whole group of Primates of which he is but one member. In so doing, we can, as it were, note the trends of evolutionary development which became manifested in the early generalized Primates and, by following them up, recognize which particular trends led to the line of evolution which culminated in Man. In other words, we are enabled to study the origin of Man in prospect rather than retrospect. This method of approach is likely to lead to a more accurate assessment of the real zoological position of Man than a method which relies (in the main) on a direct comparison with those modern apes which most closely resemble him in anatomical structure. For it needs to be emphasized that all existing Primates are but the terminal products of a number of collateral lines of descent, many of which have evidently passed through a long period of evolutionary independence. A general close structural resemblance is not necessarily indicative of derivation from a common ancestor at a relatively recent date.

It is the aim of this book to review the comparative anatomical and palæontological evidence which forms the essential basis of speculation on the phylogenesis of the Primates. By a detailed consideration of this evidence, it is possible to draw fairly definite conclusions regarding the precise interrelationship of the various subdivisions of the Primates, to recognize the numerous divergent trends of evolution which have led many of them off the main evolutionary stem, to discriminate between structural resemblances which betoken a real and close affinity and those which are the expression of evolutionary parallelism, and thus to build up a clear conception of the steps which have led to the emergence of existing species, including Man.

The problem of the evolution of the Primates is clearly of the greatest interest to all those who concern themselves with the study of Mankind. It is hoped that this book will form an introduction to this study from the morphological aspect, and that it will be particularly serviceable to those specialists who are compelled to limit their researches to one anatomical system, or to the structure of the human body without particular reference to comparative anatomy.

The study of the problems with which this book deals necessitates a very considerable and detailed knowledge of the anatomy of living Primates. Only in recent years has this information become available in sufficient quantity, and much of it is the result of investigations by anatomists in this country. This is due in very large part to the energy and guidance of my friend Professor Elliot Smith, who, more than anyone else, has recognized the importance of this knowledge, and has not only directed researches to this end in his own laboratory but has also inspired his colleagues to pursue similar studies. Every zoologist who is concerned with the study of the Primates will deem it appropriate that this volume should be dedicated to him.

W. E. Le G. C.

St. Thomas's Hospital,
          London.
    February, 1934.

CONTENTS

CHAPTER   PAGE

I. INTRODUCTION - - - - - - 1

II. THE PRIMATES : THEIR DISTRIBUTION IN SPACE AND TIME   19

III. THE EVIDENCE OF THE SKULL - - - - 40

IV. THE EVIDENCE OF THE TEETH - - - - 69

V. THE EVIDENCE OF THE LIMBS - - - - 103

VI. THE EVIDENCE OF THE BRAIN - - - - 141

VII. THE EVIDENCE OF THE SPECIAL SENSES - - - 173

VIII. THE EVIDENCE OF THE DIGESTIVE SYSTEM - - 197

IX. THE EVIDENCE OF THE REPRODUCTIVE SYSTEM - - 209

X. THE RELATION OF THE TREE-SHREWS TO THE PRIMATES - 222

XI. THE EVOLUTIONARY RADIATIONS OF THE PRIMATES - 253

INDEX - - - - - - - 289

LIST OF ILLUSTRATIONS

FIGURE   PAGE

1. THE EVOLUTIONARY RADIATIONS OF MAMMALS - - 21

2. YOUNG GIBBON - - - - - - 24

3. PIG-TAILED MACAQUE MONKEY - - - - 28

4. YOUNG TARSIER - - - - - - 33

5. DWARF LEMUR - - - - - - 35

6. THE BASICRANIAL AXIS - - - - - 41

7. SKULLS OF RECENT LEMURS - - - - - 43

8. ORBITO-TEMPORAL REGION OF THE SKULL - - - 45

9. TYMPANIC REGION OF THE SKULL - - - - 46

10. MANDIBLE OF PRIMATES - - - - - 48

11. SKULLS OF FOSSIL LEMURS - - - - - 49

12. SKULL OF *Stehlinella* AND *Chiromys* - - - - 52

13. BASE OF SKULL IN THE PRIMATES - - - - 57

14. SKULLS OF TARSIOIDEA - - - - - 58

15. SKULL OF MARMOSET AND GIBBON - - - - 62

16. GENERALIZED MAMMALIAN DENTITION - - - 71

17. EVOLUTION OF CUSPS OF UPPER MOLARS - - - 74

18. EVOLUTION OF CUSPS OF UPPER MOLARS - - - 75

19. EVOLUTION OF CUSPS OF LOWER MOLARS - - - 76

20. DENTITION OF LEMUROIDEA - - - - - 78

21. DENTITION OF LEMUROIDEA - - - - - 81

22. DENTITION OF TARSIOIDEA - - - - - 88

23. DENTITION OF *Carpolestes* - - - - - 89

xiii

FIGURE						PAGE
24. Dentition of Tarsioidea	-	-	-	-	-	90
25. Dentition of Monkeys	-	-	-	-	-	93
26. Dentition of Monkeys	-	-	-	-	-	95
27. Dentition of *Parapithecus*	-	-	-	-	-	96
28. Upper Limb Skeleton of *Notharctus*		-	-	-	106
29. Manus and Pes of Primates	-	-	-	-	109
30. Carpal Skeleton	-	-	-	-	-	110
31. Pelvic Skeleton	-	-	-	-	-	111
32. Lower Limb Skeleton of *Notharctus*		-	-	-	112
33. Tarsal Skeleton	-	-	-	-	-	114
34. Upper Limb Skeleton of *Tarsius*	-	-	-	-	116
35. Upper Limb Skeleton of *Tarsius*	-	-	-	-	117
36. Cross Sections of the Ilium	-	-	-	-	118
37. Lower Limb Skeleton of *Tarsius*	-	-	-	-	119
38. Upper Limb Skeleton of the Marmoset	-	-	-	121
39. Upper Limb Skeleton of the Marmoset	-	-	-	122
40. Manus and Pes of the Marmoset	-	-	-	-	124
41. Femur of the Marmoset	-	-	-	-	128
42. Lower Limb Skeleton of the Marmoset	-	-	-	128
43. Brain of *Centetes*	-	-	-	-	-	142
44. Brain of *Microcebus*	-	-	-	-	146
45. Formation of the Sylvian Fissure	-	-	-	147
46. Cortical Areas	-	-	-	-	-	148
47. Lateral Geniculate Body	-	-	-	-	150
48. Brain of *Adapis parisiensis*	-	-	-	-	152
49. Cerebral Sulci in a Lemur and a Monkey	-	-	154
50. Cerebral Hemisphere of the Aye-Aye	-	-	-	155

FIGURE   PAGE

51. Brain of *Tarsius*   -   -   -   -   -   - 157

52. Brain of *Tarsius*   -   -   -   -   -   - 158

53. Lateral Geniculate Body of *Tarsius*   -   -   - 159

54. Brain of the Marmoset   -   -   -   -   - 164

55. Cerebral   Hemisphere   of   a   Platyrrhine   and
         Catarrhine Monkey   -   -   -   - 165

56. Schema of Lateral Geniculate Body in Primates   -   - 167

57. Rhinarium   -   -   -   -   -   -   - 174

58. Face of *Tarsius*   -   -   -   -   -   - 175

59. Face of a Platyrrhine and Catarrhine Monkey   -   - 176

60. Nasal Cavity and Turbinate Processes   -   -   - 177

61. Section of Nasal Cavity   -   -   -   -   - 179

62. External Ear   -   -   -   -   -   - 190

63. Tongue   -   -   -   -   -   -   - 193

64. Stomach and Adjacent Viscera of *Tarsius*   -   - 198

65. Rotation of the Gut   -   -   -   -   - 199

66. Colon Pattern   -   -   -   -   -   - 200

67. Ileo-Cæcal Region   -   -   -   -   -   - 204

68. Young Tree-shrew   -   -   -   -   -   - 223

69. Skull of *Ptilocercus* and *Microcebus*   -   -   - 224

70. Skull of *Anagale*   -   -   -   -   -   - 226

71. Carpus of *Ptilocercus*   -   -   -   -   - 228

72. Lower Limb Skeleton of *Anagale*   -   -   - 229

73. Foot Skeleton of *Ptilocercus*   -   -   -   - 230

74. Dentition of *Ptilocercus*   -   -   -   -   - 232

75. Brain of *Tupaia*   -   -   -   -   -   - 234

76. Transverse Sections of Brains of *Centetes*, *Tupaia* and
         Microcebus   -   -   -   -   -   - 235

FIGURE | PAGE

77. STOMACH AND ADJACENT VISCERA OF *Ptilocercus* - - 240

78. INTESTINAL TRACT OF *Ptilocercus* - - - - 240

79. MALE EXTERNAL GENITALIA OF *Ptilocercus* - - - 241

80. EXTERNAL GENITALIA OF *Tupaia* AND *Tarsius* - - 242

81. MALE URETHRA OF *Ptilocercus* - - - - 243

82. FEMALE EXTERNAL GENITALIA OF *Ptilocercus* - - - 244

83. FEMALE INTERNAL GENITALIA OF *Ptilocercus* - - - 244

84. EXTERNAL EAR OF *Ptilocercus* AND *Tupaia* - - - 245

85. TONGUE OF *Ptilocercus* - - - - - 246

86. FOOT MUSCULATURE OF *Ptilocercus* - - - - 247

87. EVOLUTIONARY RADIATIONS OF THE LEMUROIDEA - - 257

88. EVOLUTIONARY RADIATIONS OF THE TARSIOIDEA - - 266

89. EVOLUTIONARY RADIATIONS OF THE ANTHROPOIDEA - 275

EARLY FORERUNNERS OF MAN

CHAPTER I

INTRODUCTION

THE general conception of the evolution of higher forms of life from lower forms with a simpler organization may be regarded as fairly established in the minds of biologists and, indeed, in the minds of most educated people. The evidence in favour of such an hypothesis has been accumulated in great quantity since the time of Darwin, and no alternative interpretation of this evidence has ever been offered which is in any way convincing. Within the limits of the Primates, the general scheme of evolutionary differentiation which ultimately led to the emergence of Man himself has been discussed by many authorities. This problem clearly has the greatest interest for those who are concerned with the question of human origins. While, however, we may accept the thesis of Man's descent from " lower " forms of life, there is by no means a consensus of opinion among biologists as to the precise route by which the human family arrived at its present status, or what may have been the real nature of its immediate progenitor.

It is certain that the various forms grouped together in the Order of Primates represent divergent modifications of a common ancestral type, and that the latter arose from a basal generalized stock which also provided the foundation for the development of other mammalian Orders. From such an ancestral type there diverged in early geological times a number of separate groups which, while all coming under the category of Primates by reason of a common plan of organiza-

I

tion, incorporated tendencies for evolution along their own particular lines, culminating in the development of distinctive structural characters. From one of these subdivisions, whether relatively early or relatively late, the progenitors of Man arose. The question of the precise position which Man occupies in the evolutionary development of the Primates can only be satisfactorily approached by considering the Order as a whole from the phylogenetic standpoint. It is necessary to trace the origin of the Primates to their evolutionary source and to mark out as far as possible the stages at which divergent trends of evolution became manifested within the group. The construction of genealogical trees and speculations regarding the phylogenetic origin of Man are common enough forms of popular scientific exposition, and, because of the subject with which they deal, they naturally attract a considerable amount of interest. But the origin of Man forms only a small item in the more comprehensive problem of the development of the whole group of which he represents a solitary family. It is with this major issue that we are concerned in the present instance.

By reference to anatomical and palæontological data the nature of the genetic relationships between different subdivisions of the Primates may be determined and its bearing on their evolutionary history discussed. It is true that the ultimate solution of these problems rests entirely on palæontological studies, for only when a much more complete geological record is available will it be possible to attain to any finality. But, although the palæontological sequence is still far from complete, sufficient data have been accumulated in the study of living forms and fossil remains from which it is possible to speculate with the probability of a fair degree of accuracy on the phylogenetic history of the Primates. It is obvious, of course, that such speculations, if they are to be removed from the realm of mere guesswork, must be based not only on adequate data, but also on established principles by which the data may be correctly interpreted. But, curious to say, it is seldom that phylogenetic speculators enunciate clearly the principles on which they rely.

Let us note briefly the nature of the available data and their general significance. These data may be derived from comparative anatomy, palæontology and embryology. We may first consider the kind of evidence which embryology and palæontology have to offer, not because here the evidence at present available is most abundant, but because it is conceivable that in the study of these branches of biological science records of actual phylogenetic routes may be found.

Embryological evidence of evolutionary paths has not proved of great scientific value. Unhappily, however, it has often been drawn upon with a too rash assurance to give verisimilitude to theories of one kind or another. The proposition that ontogeny recapitulates phylogeny is no doubt true in a very general way, but the recapitulation is not so specific that it may form a basis of argument for any particular theory of evolutionary descent unless the evidence is of a very positive nature. Because the foot of the human embryo assumes the characteristic human proportions very early in embryonic development, this cannot be seriously regarded as demonstrating that the human type of foot is phylogenetically extremely ancient. Again, because the arms of the human embryo may be slightly longer in proportion to the legs than is the case in the human adult, this is in no way an indication that Man has been evolved from a gorilloid ancestor with long brachiating arms and shrunken hind-limbs. And yet this type of argument is not uncommonly employed by those who advance theories regarding the origin of Man. There is, as a matter of fact, abundant evidence from the study of the development of the human body to suggest that Man evolved initially from a primitive and generalized mammalian ancestor, but surprisingly little to indicate the nature of his immediate progenitor.

In regard to palæontological evidence, it is clear that if in the same geographical region large numbers of fossil remains are found in geological strata of different and successive ages, and if they exhibit a graded series of structural changes linking up an ancient form X with a modern form Y, then here we have presumptive evidence of the evolutionary route by which Y may have been derived in the process of time from X. Unfortunately, palæontological records even approaching such completeness are very rare, and they can hardly be said to exist so far as the evolution of the Primates is concerned.

The major part of the evidence upon which zoologists have perforce to depend for the reconstruction of the phylogenetic lines of the Primates is therefore derived from the study of the comparative anatomy of existing forms and of such fossil forms as have hitherto been discovered. It is necessary now to examine the significance of this evidence in some detail in order to test its validity.

Briefly speaking, degrees of genetic affinity are assessed by noting degrees of resemblance in anatomical details. Having defined a group of animals which are believed on these grounds to be closely related, it is possible by a sort of mental triangulation to postulate the nature of the ancestral form from which it originated in the evolutionary sense. Bearing in mind the variously modified characters of the members of such a natural group, and having a general conception of the direction of evolutionary development from the generalized to the specialized, a morphological common denominator or prototype is hypothecated from which it is readily comprehensible that the diverse modifications in the existing group may have been derived. This method of reasoning involves several considerations. First of all, the taxonomic significance of different anatomical characters requires attention. It is generally believed that in estimating genetic relationships much more stress should be laid on some characters than on others. Thus, anatomical characters have been divided by systematists into two categories—adaptive and non-adaptive. The former bear an obvious and direct relation to some functional demand which itself is dependent upon the particular circumstances of the environment. These characters are regarded as of less taxonomic value in so far as it is supposed that they may be developed independently in unrelated forms which happen to be subjected to similar environmental forces. A non-adaptive character, on the other hand, is one which is presumed to bear no direct relation to function and to be, therefore, an incidental feature in the structural composition of an animal which is of no economic significance to its owner. Such a character, which may persist in the face of gross changes in the environment while other characters of less morphological stability show striking modifications, is obviously of particular significance for the systematist. The difficulty, however, is to decide when an anatomical feature is non-adaptive. Often it may be presumed to belong to this category merely because no functional significance happens to have been discovered for it. It may not infrequently be the case that a feature which is regarded by one zoologist as essentially non-adaptive (and upon which he lays considerable stress in estimating genetic affinities) may be held by another to bear a direct relation to functional requirements. It is clear, therefore, that even presumed non-adaptive characters must be used with the greatest discretion in the determination of genetic relationship.

For the latter purpose perhaps the greatest value should be attached to those anatomical structures which appear to subserve identical functions and at least superficially resemble one another, but which may be constructed on a variety of plans and from a variable combination of elementary parts in different animals. A complete identity in regard to such a structure argues most strongly for real genetic affinity, while a fundamental difference (which may not be easy to detect except on very close analysis) is equally suggestive of an evolutionary parallelism which has led to resemblances which are superficial only.

The details of the cusps of the molar teeth may differ quite considerably in a group of animals in which the diet offers no corresponding contrast to the closest observation.   Similar cusp patterns may be found in two Primates which at first sight appear to be closely related.   But on analysis it is perhaps found that the similarity has been arrived at in different ways as the result of the different mode of development of one of the cusps.   Dental contrasts such as these indicate that the relationships may be considerably more distant than a cursory examination would suggest.   The converse may be exemplified by reference to the curious colon pattern which is found in lemurs, and which is equally developed in animals whose diet is very different.

The bony tympanic chamber of the skull forms an essential part of the mechanism of hearing and apparently meets identical functional demands in different mammals.   Yet the bony elements which are utilized in the construction of this chamber are by no means the same in all mammals, or even in mammals which have been supposed to be closely related forms.   Such differences supply the most cogent evidence for assessing taxonomic position and far outweigh any number of anatomical resemblances which conceivably owe their existence to evolutionary parallelism.   In other words, it may be said that in the evaluation of genetic affinities anatomical differences are more important as negative evidence than anatomical resemblances are as positive evidence.   It becomes apparent that if this thesis is carried to a logical conclusion, it will necessarily demand a much greater scope for the phenomenon of parallelism or convergence in evolution than has usually been conceded by evolutionists.   The fact is that the minute and detailed researches which have been carried out by comparative anatomists in recent years have made it certain that parallelism in evolutionary development has been proceeding on a large scale and is no longer to be regarded as an incidental curiosity which has occurred sporadically in the course of evolution.   Indeed, it is hardly possible for those who are not comparative anatomists to realize the fundamental part which this phenomenon has played in the evolutionary process.

Actually there is no *a priori* reason why parallelism should not be a common phenomenon. Organisms which have been derived at one time or another from the same ancestral stock must presumably be endowed with the identical potentialities which they have inherited in common, and it is only to be anticipated that they will tend to react in the same way under the influence of similar environmental conditions. The reluctance to accept such a proposition appears largely to be associated with the conception of evolution as a haphazard process depending upon the natural selection of a succession of minute chance variations and thus leading by an incredible number of what appear to be accidental progressive steps to the completion of some organ of great complexity. On this idea it would seem impossible, for instance, that the elaborate brains of the large Old World monkeys and the large New World monkeys (which are remarkably similar even in histological details) could have been derived independently from a small and simple brain of the type seen in the modern marmoset. And yet, that this is precisely what has occurred is evident if all the facts of the case are taken into account. Moreover, if parallelism is capable of expressing itself to this extent, it would be difficult to place any limit on its possibilities. In this case it is clear that close anatomical resemblances are not necessarily indicative of close genetic affinity in the ordinarily accepted sense of the phrase. On the other hand, even resemblances based on parallelism may be interpreted as indicating an *ultimate* genetic relationship in so far as they are regarded as an expression of the fundamental principle enunciated by H. F. Osborn, that the same results always tend to appear independently in descendants of the same ancestors. Certainly, as a working hypothesis, it is justifiable to assume that organisms which show a preponderating resemblance to each other anatomically are genetically related forms, unless some flagrant discrepancy exists in one or more features such as could only be explained morphologically by supposing a long period of independent evolution from an ancestral form of a much more primitive type.

It follows from this discussion that any attempt to construct a classification of animals by reference to one organ or one anatomical system only may lead to the most fallacious results. The evidence of external characters may be flatly contradicted by the evidence of the teeth; the structure of the alimentary tract may suggest close affinities in animals in which the brain shows quite clearly that they cannot possibly be closely related, and so forth. *All anatomical characters must be taken into account in assessing affinities.* If the closest scrutiny of the entire structural organization of two animals reveals no significant contrast between them, it may be inferred that they are closely related forms which have been derived at no very distant epoch from a common ancestral type.

That the general direction of evolution has been from the primitive generalized type with a relatively simple organization to the specialized or "higher" type with a complex organization is a fundamental element of the evolutionary hypothesis. By this is understood that the higher type of animal is more elaborately organized in so far as it is more subtly adapted to its particular environment and more capable of dealing with the accidents and emergencies which may arise in connection with its mode of life. The nervous system, for instance, shows a greater intricacy in its structure and connections. It does not mean necessarily that the morphological elements of which the different parts of the animal are composed are in every case more complexly arranged. Thus the progressive evolution of the skull in vertebrates has been marked, in general, by a progressive simplification of the adult structure, for in the higher vertebrates fewer separate elements take part in its composition than in the lower. The dentition of higher types may become simpler than those of lower and ancestral types as the result of the suppression of one or more teeth.

In order to define the evolutionary status of any particular member of a group of mammals, it is clear that we must have some conception of the nature of the generalized prototype which represents the ancestral basis from which this group originated. For this purpose, we must be able to distinguish between features which are essentially primitive and those which are specialized. This can be done (1) by reference to the existing *échelle des êtres*, with a consideration of the anatomical structure of those forms which are on the whole the most simply organized and occupy the most lowly position in a scale of increasing elaboration; (2) by reference to the early fossil representatives of the group, which appear to form the ancestral basis for its subsequent evolutionary development; (3) by detailed anatomical study of the anatomical feature concerned; and (4) by reference to embryology. Thus, for instance, pentadactyly is judged to be a primitive and generalized condition in mammals which has in some cases been replaced by specialization depending upon the loss of one or more digits. For in living mammals of a simple organization and in Reptilia (from which mammals were originally derived) pentadactyly is the general rule; palæontology has demonstrated that in the early forerunners of mammals which to-day have less than five digits, pentadactyly was also a characteristic feature; and, lastly, a detailed study of mammals with less than five digits reveals in the adult vestigial remains of those which have been lost during the course of evolution or, in the embryo, transient traces of these vanished structures.

In many cases the distinction between a primitive and a specialized trait is so obvious that it hardly requires discussion. The evolutionary history of the brain has clearly been marked by a progressive elaboration and increasing complexity of its higher functional levels, especially of the cerebral hemispheres, and in comparing two brains of an evolutionary series there can never be any doubt as to which is the more primitive and which the more advanced. In some instances, however, there may be some difficulty in determining which of two structural modifications is the more generalized, for only after protracted research and prolonged discussion is it possible to affirm the direction of evolution. This is the case, for example, with certain features of the dentition. The molar teeth in mammals have a variable number of cusps. One theory advocates that in the generalized condition there were a large number of cusps, and that the less complicated molars of many existing forms are the result of secondary simplification following on the suppression of some of the cusps. This is the multitubercular theory. Another theory maintains that the generalized mammalian molar had only three cusps and that specialization has proceeded rather by the addition of a varying number of new cusps. This is the tritubercular theory. It is not possible here to enter into a detailed review of these theories ; it must suffice to note that, although there still remain some adherents of the first theory, the accumulation of evidence undoubtedly weighs vastly in favour of the tritubercular theory.* It may be assumed, therefore, that the many-cusped teeth are the result of specialization (even if they occur in some geologically early forms), while tritubercular molars represent a generalized condition (even if they persist to the present day, as in the small Primate, *Tarsius*).

It may be accepted, in general, that as a result of the morphological researches of the last half-century we are to-day in a fairly safe position to discriminate between generalized and specialized features. It must be noted, however, that a specialized character may undergo secondary degeneration resulting in a structure which simulates a more primitive phase of evolutionary development, and in some instances there may be doubt as to whether we are dealing with a truly primitive feature or a secondary regression. But it is probable

* Within the limits of the Primates, the palæontological sequence represented by the fossil remains of the primitive lemuroid *Pelycodus*, from early tertiary deposits of Wyoming, provides very convincing evidence in favour of the tritubercular theory. Starting with *P. ralstoni* and ending with *P. tutus* (and leading up to the genus *Notharctus*), the changes postulated by this theory are illustrated in a remarkable way.

that in the vast majority, if not in all, of these cases a close scrutiny will reveal the true state of affairs. For example, the canine tooth in modern Man is a small incisiform tooth which shows very little differentiation in comparison with the adjacent teeth. This appears at first sight to be considerably more primitive than the condition in the anthropoid apes in which the canines are specialized as large and powerful pointed teeth. But in Man the long root of the canine and its lateness in eruption—as well as the evidence of fossil forms—betray its real nature as a tooth which has undergone at least some degree of reduction. In the brains of apes and monkeys there is a very distinctive sulcus which was thought to be absent altogether in the human brain. It is often called the " Äffenspalte " or the simian sulcus. A superficial study might suggest that it is a highly specialized trait of the ape's brain which has been avoided in the human brain, and that the latter is in this respect more primitive and generalized. But Elliot Smith demonstrated many years ago that this sulcus is also present in the human brain, that it is extremely variable (a characteristic feature of degenerative structures in general), and that it may be as well developed relatively in brains of certain races as it is in the anthropoid apes. It may be inferred, therefore, that the simian sulcus has undergone a retrogression in Man and that the human brain is thereby less primitive than the ape's brain.

Conversely it is legitimate to assume that any character which has the appearance of being primitive and generalized is in fact *really* primitive and generalized—so long as it displays no details in its construction, development, or evolutionary history which point to a secondary simplification or retrogression.

It is important to note that an extremely primitive condition in one anatomical system may not infrequently be associated with a high degree of specialization in another in the same animal. Such a combination has sometimes led to doubt in the minds of certain anatomists whether the primitive features are truly primitive. Thus, for instance, it has been argued that the placentation of lemurs cannot be as primitive as it appears because lemurs are in many respects specialized creatures. Reasoning of this kind is quite invalid. It is well known that an archaic simplicity of structure may be retained in one part of the body at the cost of extreme specialization affecting other parts. Man, for instance, has been able to preserve many astonishingly generalized anatomical characters in virtue of the very advanced development of his brain. Indeed, it would be not unreasonable to argue that a seemingly primitive trait is the more likely to be truly primitive if it *is* associated with extremes of specialization elsewhere.

These considerations of the distinction between generalized and specialized are closely bound up with the principle which postulates the Irreversibility of Evolution, to which attention may now be turned. This is a generally accepted principle, and it is usually relied upon in attempts at the reconstruction of phylogenetic histories. Indeed, it is a principle which must be relied upon in the absence of an approximately complete palæontological record if any conclusions at all are to be reached. The evidence for this Law of Irreversibility is necessarily of an indirect nature, but, such as it is, it certainly indicates that a positive reversal, the regaining of characters which have once been lost, is probably so rare that it may be discounted altogether in phylogenetic speculations unless there is the strongest evidence for its presumption.

One authority has within recent years stated as his opinion that " of all the so-called ' laws ' of evolution that have been formulated, that of the ' irreversibility of evolution ' has perhaps a more general application than any."* This idea of irreversibility, indeed, is implicit in the arguments of probably every evolutionist. The conception of evolutionary trends with progressive specializations in various directions, accompanied by loss of evolutionary plasticity, is inherent in

* C. Tate Regan, *Proc. Zool. Soc.*, 1924, p. 175.

all attempts to unravel lines of descent from the evidence of comparative anatomy. Multitudes of supposed missing links have been discarded as such after more careful scrutiny because they have been found to possess one or more characters which are regarded as less primitive than the corresponding characters of the forms to which they were at one time assumed to have given rise.

Irreversibility is an integral part of the conception that the process of evolution has been by no means a haphazard affair, but an orderly, regular and progressive sequence of changes dependent upon definite laws the nature of which is not yet properly understood. What may be termed negative reversal—the loss of a character which has previously been developed—is a common enough happening in evolution, but, as pointed out above, the degenerative nature of such a process is usually betrayed by an intimate examination of the part concerned. A number of muscles which presumably became differentiated in a primitive mammalian phase of evolution have disappeared in Man, but their history is evidenced by fibrous vestiges or by their occasional appearance as " atavisms " in individual cases. Instances of a positive reversal have been adduced as rare curiosities of phylogenesis, but even these apparent exceptions are open to an alternative interpretation.

From the established principle of the irreversibility of evolution certain corollaries arise which are of the utmost importance in tracing lines of phylogenetic development. Thus it may be stated that the ancestral type from which a given natural group of animals may have arisen must have been at least as primitive in all its characters as the most primitive member of that group. Conversely, no anatomical feature of any animal can be more primitive in the sum of its structural details than the corresponding feature of its evolutionary progenitor. These may seem obvious truisms, but attention is frequently not given to them. The most primitive type of brain seen in the Anthropoidea is that of the marmoset.

It may be inferred, therefore, that the evolutionary progenitor of the monkeys must have possessed a brain at least as primitive. An extinct group of lemurs which was more specialized in dentition and cranial characters than the generality of the modern monkeys could hardly have provided a foundation for the evolutionary development of the Anthropoidea. By reference to a number of anatomical systems it is in this way possible to construct a mental picture of the general make-up of ancestral types, to differentiate those fossil types which represent aberrant or divergent offshoots of the main line of descent from those which may occupy a position on or close to the main stem, and to gain a fairly clear conception of the scheme of evolution of any particular group of animals.

In some discussions on phylogenesis, and especially the phylogenesis of Man, confusion has been introduced by the adoption of an uncertain terminology in reference to the mammalian groups concerned. It is essential that the nomenclature should be defined as clearly as possible if profitless controversy is to be avoided. Obviously, from the evolutionary point of view, it is impossible to give a precise definition of a natural group in terms of morphological characters which have been acquired in their complete form. The Primates include all those types which have existed since the immediate precursors of the Order as a whole became separated from the basal mammalian stock which also gave rise to other Orders. Thus the earliest Primates would have shown none of the characteristic features of the fully differentiated genera—such as a large brain, reduction of jaws, and so forth. It is therefore not feasible to define a group by the characters which are found in its existing representatives, the end-products of long lines of evolution.

A classification must necessarily be labile so that it can be readjusted to keep pace with palæontological discovery. The only satisfactory method of defining a natural zoological division is by reference to its phylogenetic differentiation and in terms of the evolutionary tendencies which it has manifested since its primary origin in common with other divisions from a basal ancestral stock. This method of definition has been adopted in the last chapter of this book. Such definitions or " diagnoses " have the advantage of being much less static and thus less artificial than the method of classification of animals usually employed.

As examples of the difficulties to which nomenclature has led in evolutionary discussion, the following instances may be noted. One authority may vigorously deny that Man is derived from an anthropoid ape, because by the term " anthropoid ape " he means an arboreal animal showing certain adaptive specializations such as are found in modern apes and which would presumably be absent in any form ancestral to Man. This conception of " anthropoid ape " is based on a common textbook definition and, so far as it goes, the argument is quite valid. Another authority, who accepts the anthropoid ape ancestry of Man, obviously uses the term " anthropoid ape " with a much wider connotation. In so far as the term may be taken to refer to a group of Primates which occupy the same evolutionary status as the modern apes in regard to brain development, general proportions of the skull, etc., but in which certain specializations such as the atrophy of the thumb need not be essential characters, this second view is also correct. As we shall see, there is certainly evidence that the Hominidæ have been derived from an ancestral stock whose general morphological characters would legitimately allow the application of the descriptive term " anthropoid ape " even though it would not have presented some of the divergent modifications shown in the living apes. This means that the term is used here with a wider application than it is generally accorded by systematists who appear to confine their attention mainly to living mammals. In other words, recognition is taken of phylogenetic considerations in defining a natural group such as the anthropoid apes or Anthropomorpha. The " anthropoid ape stock " should be taken to include not only the end-products of evolu-

tionary lines such as the modern apes, but also all those intermediate types which have come into existence since the
progenitors of the group first became segregated from the
progenitors of the Catarrhine monkeys. It is evident, from
this point of view, that the earliest representatives of the
Anthropomorpha would not have manifested the morphological specializations of the modern apes, and some of them,
in avoiding these specializations, might well have provided
a basis for the evolution of the Hominidæ.

Again, it has been argued that the whole stock of the Old
World monkeys is excluded from any share in human ancestry.
In so far as the whole group of Old World or Catarrhine
monkeys is technically defined as being characterized by certain
specializations (such as bilophodont molars and ischial callosities) not to be found in Man—that is to say, in so far as
these features are accepted as *essential* components in the make-
up of an animal which we can call an " Old World monkey "—
it cannot be truly said that Man has ever evolved from such
a type. But if an animal were discovered which in general
characters (*e.g.*, dental formula, cranial features, etc.) con-
formed with the known Old World monkeys, but which lacked
the few aberrant modifications which characterize the latter,
it would almost certainly be called a Catarrhine monkey,
and the definition of the Catarrhines would presumably be
extended to accommodate it. The fossil *Parapithecus*—known
from a mandible and teeth—is generally regarded (on the
basis of dental characters) as a very primitive anthropoid ape.
Were the whole of its anatomy available for study, it might
be deemed more appropriate to include it with the Catarrhine
monkeys even though in certain features it was much less
specialized than the living representatives of this group.
Moreover, it seems that this form may readily provide a
morphological basis for the evolutionary origin of the apes
and Man.

As another analogy, we may suppose that if some pre-
Columbian zoologist in Europe had made a study of the

Primates then available to him, he would have included in his definition of monkeys the possession of two premolars and of ischial callosities. But the subsequent discovery in America of very similar animals with three premolars and no callosities would not have prevented him from calling them true monkeys. It would have led him merely to realize that his first definition had clearly been too limited and to that extent did not accord with the natural grouping.

Thus it is clear that current zoological classifications require to be treated as somewhat arbitrary attempts to separate the known types into well-defined groups for the convenience of description and cataloguing. In order to reflect the realities of nature, definitions based on anatomical characters should be made as general as possible, and the larger the group the more elastic does its definition require to be.

It may seem a little superfluous to enter on this discussion on nomenclature. For the outcome of it is merely that the question whether or not Man has descended from a monkey depends on what is meant by a " monkey." It is necessary, however, to stress these points, because a too rigid conception of zoological classification has evidently been a fruitful source of misinterpretation in the past. Disputants who are engaged in similar controversies involving the question of lemuroid or tarsioid ancestry would do well to begin by defining what they understand by these terms. If Man is to be denied a pithecoid ancestry because monkeys are technically defined in systematic textbooks as possessing certain specialized features characteristic of their *modern* representatives, then on the same grounds a tarsioid or lemuroid ancestry must obviously be excluded also.

In the ensuing chapters the various anatomical systems of the Primates will be dealt with separately in an attempt to see how far they may provide evidence bearing on Primate phylogeny. In the last chapter this evidence is considered

2

as a whole. Some anatomical systems, such as the vascular and respiratory systems, have not been given separate treatment, for the reason that, so far as comparative details concerning them are at present available, they provide no very significant evidence pertinent to the main thesis under discussion.

CHAPTER II

THE PRIMATES: THEIR DISTRIBUTION IN SPACE AND TIME

THE origin of mammals from a reptilian ancestry is accepted by all biologists as an established fact. Palæontological investigation has revealed the existence at the end of the Palæozoic epoch (the first of the three great subdivisions of geological time) of a large number of mammal-like or Theromorph reptiles which provide remarkable structural transitions between true reptiles and true mammals. During the Permian age—which marks the last phase of the Palæozoic epoch—certain small Theromorphs flourished in South Africa which very probably represent the ancestral group from which mammals were initially derived. It is not until the middle of the Mesozoic or Secondary epoch, however, that undoubted mammals become evident in the geological record. During the central phase of this epoch—the Jurassic age—mammals became relatively abundant and widespread, though it seems that most of the mammalian fauna of this time was destined to become extinct.

Dr. G. G. Simpson has dealt extensively in two monographs with the Jurassic mammals of the Old and New Worlds, and he suggests that, of the various large groups which he has been able to define, one—the Pantotheria—represents the stock which ultimately gave rise to the placental mammals.[6] Other great groups such as the Multituberculata, Triconodonta and Symmetrodonta, were already extinct in the early part of the Tertiary epoch. It is at least fairly certain that at the very base of the mammalian stem, and quite soon after its emergence from the ancestral Theromorph stock, there separated out three great radiations which persist to-day as the Monotremata (the egg-laying mammals of Australasia), the Marsupialia (or pouched mammals), and the Placentalia. The latter are distinguished (*inter alia*) by the fact that their young are nourished in the uterus by a highly elaborated (allantoic) placenta and are born in a relatively mature condition. In the Marsupials, on the other hand, only an elementary form of placenta is developed during the intra-uterine life of the embryo, while the young are born still in an embryonic stage and continue their development in the marsupial pouch. These three sub-classes represent broadly three levels of evolutionary development of mammals which have been termed Prototheria, Metatheria, and Eutheria. There is abundant reason to suppose that the placental mammals (which represent the eutherian stage) have in their phylogenetic history passed through metatherian and proto-therian stages, but, as Simpson has been careful to emphasize, this does not necessarily mean that they were derived succes-sively from the Marsupialia and the Monotremata, which are surviving representatives of these stages. The two latter sub-classes are to be regarded as side-branches which have preserved primitive characters to a much greater extent than the Placentalia, but which have also undergone aberrant specialization peculiar to themselves.

By the beginning of the Tertiary epoch the placental mammals had become differentiated into a number of separate Orders most of which have persisted to the present day. In Fig. 1 is reproduced a diagram constructed by the late Dr. W. D. Matthew[4] to represent the progressive differentia-tion of the Orders of mammals during the Mesozoic and Tertiary epochs.* We are concerned in this thesis entirely with the Order of Primates.

* For the convenience of those who are not familiar with the divisions of geological time, it may be noted that the Mesozoic epoch is sub-divided into three phases: the Triassic, the Jurassic, and Cretaceous, of which the last is the most recent. The Tertiary epoch is likewise

Probably in the middle of the Cretaceous period the ancestral or basal Primate stock became segregated from the common eutherian stock which also gave rise to the other

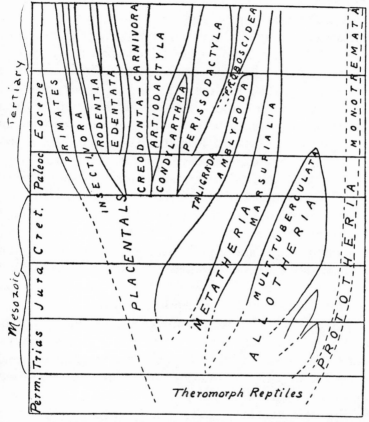

FIG. 1.—A SCHEMA SHOWING THE PROGRESSIVE DIFFERENTIATION OF THE ORDERS OF MAMMALS DURING THE MESOZOIC AND TERTIARY EPOCHS. (W. D. Matthew, *P.Z.S.*, 1928.)

placental Orders, and from that date it entered upon an independent evolutionary history which culminated in the differentiation of a number of well-defined groups or subdivisions.

divided into the following periods from its commencement: Palæocene, Eocene, Oligocene, Miocene, Pliocene, Pleistocene, and Recent. The last two periods are sometimes grouped together as the Quaternary epoch.

The earliest Primates must clearly have been of an extremely primitive character, differing imperceptibly from the earliest representatives of the other Orders. They were at first characterized only by the fact that they incorporated potentialities for evolutionary development along certain definite lines, and they eventually gave rise to a group of mammals all of which are to be distinguished by a complex of structural features which they possess in common. These structural features form the basis for the definition of the Primates as contrasted with other Orders. Although they will be dealt with in detail in the main part of this book, it is convenient at this stage to note that the Primates have manifested a prevailing tendency towards the development of relatively large and complicated brains, the elaboration of their visual powers, the reduction and atrophy of olfactory mechanisms, the shortening of the snout or muzzle, the preservation of a generalized structure in the limbs associated with a free mobility of the digits and especially the pollex and hallux, and the replacement of sharp compressed claws by flattened nails.* These characters may be correlated with their particular habitat, for, with very few exceptions, all the Primates are arboreal creatures. The exceptions are represented by certain genera which have demonstrably become adapted secondarily to a terrestrial mode of life. By reference to their external characters the majority of living Primates can be recognized as such even by the casual observer.

The Primates are most conveniently divided into three sub-orders, the Anthropoidea, Tarsioidea and Lemuroidea.[3, 8] The two latter may be included for descriptive expedience under the term Prosimiæ. Alternative and more elaborate classifications have been proposed from time to time (to some of which reference will later be made), but we are not concerned here with a detailed discussion of their merits from the point of view of the taxonomist. We will content ourselves with the commonly accepted scheme of classification

* For a more detailed definition of the Primates, see the last chapter.

for which, indeed, there is considerable justification, as the succeeding chapters will show.

In the present chapter a general survey of the different groups of living and fossil Primates is given, the main purpose of which is to render intelligible the references which are made to them in the detailed discussions of Primate anatomy which follow. The three sub-orders may be dealt with separately, and no attempt is made to give a complete list of all the known genera.

The Anthropoidea

This division includes Man, the anthropoid apes and the monkeys. As the sub-ordinal name implies, the members of the group are distinguished by their man-like appearance, which catches the attention of the most unsophisticated eye. The human resemblance of the apes and monkeys on analysis resolves itself into a few outstanding characters such as the relatively voluminous and rounded brain-case and the flatness of the face ; the position of the eyes, which look directly forwards and so appear rather close-set ; the shrunken appearance of the external ear ; the alertness and versatility of the facial expression, conditioned partly by the advanced degree of differentiation of the facial muscles ; the mobility of the lips, and especially the upper lip which is not cleft and bound down to the gums as in most lower mammals ; the absence of a naked moist surface round the nostrils ; the employment of the manus as a real "hand" for grasping purposes, combined with the functional independence of the thumb ; and the presence of flattened nails on the terminal phalanges.

At the head of this sub-order are placed the Hominidæ. Only one species of this family exists to-day, *Homo sapiens*. The phenomenal expansion of the brain in *Homo* marks the culminating point of an evolutionary tendency which is displayed to varying degrees among all the Primates. Several fossil species of the genus *Homo* have been recognized—*e.g.*, *Homo neanderthalensis* and *Homo rhodesiensis*, which parallel the modern species in the actual volume of the brain but show also aberrant or primitive features. Extinct genera may be instanced by such forms as *Pithecanthropus*, *Sinanthropus* and *Eoanthropus*. The first two are classed by some authorities as a distinct family, Pithecanthropidæ, and all are referable to the early Pleistocene or late Pliocene age. *Pithecanthropus* is represented by the famous Java remains discovered in

FIG. 2.—A YOUNG GIBBON.

From a photograph by the author. Note the excessive length of the arms, the use of the hands as hooks, and the relatively short legs.

1898 by Dubois, while *Sinanthropus* was recently found in China close to Peking. In many respects these fossils fill in the structural gap between *Homo* and the anthropoid apes in a most remarkable way, and, together with *Eoanthropus*, they bear witness to the wide geographical distribution of the Hominidæ during the later phases of the Tertiary period. The remains of *Eoanthropus* were derived from a gravel deposit in Sussex and this genus is especially noteworthy because it combines a brain-case which approximates to (but by no means closely resembles in detail) that of modern Man, with a mandible and dentition which are extraordinarily ape-like.

The anthropoid apes, which form a group called the Anthropomorpha, are represented by four living genera, Gorilla, Chimpanzee, Orang-Utan and Gibbon. The first three comprise a single family, the Simiidæ, while the gibbon is usually separated to form a distinct family, the Hylobatidæ.

The living anthropoid apes approximate structurally more closely to Man than do the monkeys in the relative size and configuration of the brain, in many details of the skull, skeleton and dentition, in the tendency to adopt a semi-erect or orthograde posture which is correlated with certain features such as the shape of the thorax and the disposition of the abdominal viscera, etc., and in the absence of a tail. On the other hand, they are pre-eminently arboreal specialists, and in association with their habit of swinging among the branches with their arms (brachiation) they have developed distinctive specializations such as an undue lengthening of the fore-limb and a varying degree of atrophy of the thumb. The modern gorilla, however, is only slightly arboreal, spending most of its life moving about in the undergrowth of the jungle and usually adopting a quadrupedal gait. But the anatomical evidence indicates clearly that this genus has only recently in its phylogenetic history abandoned brachiating habits in association with which its limbs have evidently been modified in the same way as in the other apes. The gorilla inhabits the equatorial regions of Africa and is the largest of the anthropomorphous apes. Closely related to it is the chimpanzee, which has a similar habitat, but is smaller and a much more active climber. The orang-utan is an Asiatic ape, confined to Borneo and Sumatra. It attains to a fairly large size and is exclusively arboreal, only rarely descending to the ground. These three apes may be referred to as the giant anthropoid apes in contrast to the gibbon, which is relatively a much smaller animal. The gibbons are found in the south-eastern parts of Asia, extending over a fairly wide area.   They are somewhat slenderly built, and their arms attain to a great length (Fig. 2). In a number of features they provide a structural transition between the giant anthropoid apes and the Old World monkeys, e.g., in the possession of ischial callosities and in the structure of the brain.

Fossil anthropoid apes are known in some numbers.   The earliest is perhaps *Parapithecus*, from the lower Oligocene deposits of Egypt, though the anthropoid ape status of this genus (of which only the mandible and lower dentition are known) is uncertain.   A separate family of the Anthropomorpha, Parapithecidæ, has been suggested to accommodate it. From the same deposits there has been found the jaw of *Propliopithecus*, which is believed to represent a genus quite closely related to the immediate ancestors of the recent gibbons.   *Pliopithecus* and *Paidopithex*—European genera of lower Miocene and lower Pliocene age—are still more advanced forms of anthropoid ape, and also resemble the gibbons so far as can be judged by rather fragmentary remains.   Together with *Propliopithecus*, they are included in the Hylobatidæ. Giant anthropoid apes are known to have had a considerable geographical range during Miocene times, as evidenced by the remains of *Dryopithecus* in Europe and Asia, *Palæosimia*, *Palæopithecus* and *Sivapithecus* from India, and *Australopithecus* from South Africa.   Of these forms, some have been regarded as representatives of the stock which gave rise to the Hominidæ, but with the exception of *Australopithecus* the evidence, though it may be highly suggestive, rests almost entirely on dental characters.   It is unwise, therefore, to do more than tentatively consider possibilities until the skull and skeleton of these fossils are known in some detail.

The monkeys represent the lowest stratum of the Anthropoidea which exists to-day.   Most of them are thoroughly arboreal, but many (especially the Old World monkeys) quite frequently come to the ground, and some are wholly terrestrial.   Both in the branches and on the ground they commonly assume a pronograde posture and a quadrupedal gait.   Unlike the Anthropomorpha, the fore-limbs are shorter than the hind-limbs—a primitive mammalian feature.   The majority, also, have well-developed tails.   The monkeys are divided into two clearly defined groups, the Old World or " Catarrhine " monkeys and the New World or " Platyrrhine " monkeys.   The significance of these terms (referred to in Chapter VII.) is related to the disposition of the nostrils, but this is a distinction which is not always very striking (see page 177). - The Catarrhine monkeys are distinguished sharply by the possession of conspicuous sitting pads or ischial callosities, the development of cheek pouches (which, however, may be absent), the dentition, and certain cranial characters.

It should be noted that while the Platyrrhini are equivalent to the New World monkeys, the term Catarrhini is commonly taken to include Man and the anthropoid apes as well as the Old World monkeys.   The latter are then frequently referred to as the Cynomorpha to distinguish them from the Anthropomorpha and Man.   But this classification seems to approximate rather too closely in one group the Old World monkeys and Man, in contrast to the group represented by the New World monkeys.   In this book the term " Catarrhine " will be employed entirely in adjectival form, with reference to the Old World monkeys in contradistinction to the Anthropomorpha.

The suggestion has been put forward that all the monkeys might with propriety be grouped in one special sub-order of the Primates, the Pithecoidea, in which case the sub-order Anthropoidea is reserved for the apes and Man.   Apart from the confusion to which this leads, there seems no adequate ground for separating so widely the monkeys and apes, especially as the Hylobatidæ provide in many respects a transitional stage between the two.   But it is convenient to reserve the adjective " pithecoid " to connote collectively all the monkeys (Platyrrhine and Catarrhine), and it will be so used here.   The term " anthropoid " should properly speaking refer to all the Anthropoidea. As Wood Jones[9] has pointed out, however, this word has been so often used with diverse meaning (it has even been employed to signify ape-like !) that it were better avoided. For this reason, the adjective " anthropomorphous " may be used in referring to the anthropoid apes and " human " or " hominoid " in referring to Man.

The Catarrhine monkeys comprise one family only, the Cercopithecidæ. They have a wide distribution over the

FIG. 3.—THE PIG-TAILED MACAQUE (*Macacus nemestrinus*).

From a photograph by the author, illustrating the appearance of a typical
Catarrhine monkey. Note that, unlike the anthropoid apes, the
hind-limbs are longer than the fore-limbs.

Old World. The genus *Macacus* (or *Macaca*)* is perhaps the most familiar type (Fig. 3). The macaque monkeys are rather stoutly built animals of moderate size, and their geographical range extends from North Africa to Japan. *Cercocebus*, the mangabey monkey, is restricted to Africa, and is characterized by conspicuous white upper eyelids and a long tail. It is more thoroughly arboreal than *Macacus*.

* The nomenclature of the monkeys is still hardly agreed upon. In this, and in some other instances, alternative names which are in use have been indicated.

Cercopithecus is another African genus rather similar to *Cercocebus*, but showing certain differences in the dentition. It comprises a large number of species, many of which are brightly coloured with rather unusual tints. *Erythrocebus* is a more terrestrial African form. The baboons (*Papio*) and mandrills (*Mandrillus*) inhabit rocky open country, and seem to have abandoned arboreal habits almost entirely. They are quadrupedal animals with large projecting muzzles and short tails, usually living in well-established communities. *Cynopithecus* is a genus which is found in the Celebes, and—so far as its external appearance is concerned—stands in an intermediate relation to *Macacus* and the baboons.

All the genera hitherto mentioned are grouped in one subfamily, Cercopithecinæ. In another sub-family, the Semnopithecinæ, are collected the langurs (*Semnopithecus*, *Presbytis*, etc.), the guereza monkeys (*Colobus*) and the proboscis monkey (*Nasalis*). They are distinguished as a whole by the absence of cheek-pouches, and the development of an elaborate and sacculated stomach in association with their strictly vegetarian habits. They feed mainly on leaves and the young shoots of certain plants, and in captivity are much more delicate animals than the Cercopithecinæ. The langurs are distributed over Malaysia, extending up to China. *Nasalis* is confined to Borneo and is remarkable for the development of a conspicuous fleshy proboscis, which is much larger in the male and gives the animal a most grotesque appearance. *Colobus* is an African genus of which some species develop a beautiful coat of long hair. In correlation with its specialized arboreal habits, the thumb is either very greatly reduced or may be absent altogether.

Fossil Catarrhine monkeys have been found over a wide area in the Old World. *Apidium*, from the lower Oligocene of Egypt, is regarded as a precursor of the recent Cercopithecidæ, showing some rather primitive features in its dentition. *Mœripithecus*—known by two molars from the same geological deposit—is probably closely related to *Apidium*. The oldest representative of the Cercopithecidæ which can be fairly definitely assigned to this family is *Oreopithecus* from the early Pliocene of Europe* (though Schwalbe believed this genus to be sufficiently distinctive to warrant the creation for it of a special family, the Oreopithecidæ). *Mesopithecus* is known by an almost complete skeleton from the lower Pliocene of Greece, and remains of this extinct monkey have also been discovered in deposits of a similar age in Hungary, South Russia and Persia. It is a well differentiated cercopithecid and is regarded (on its skull characters) as a representative of the Semnopithecinæ. To the same sub-family are referred *Dolichopithecus* from the Pliocene of France and *Libypithecus* from the middle Pliocene of Egypt. The modern genus *Semnopithecus* was also represented during Pliocene times in France and North India. Remains of true baboons are known from the Pliocene levels of North India (the Siwalik deposits) as well as from the Pliocene of Egypt and Algiers. Finally, *Macacus* was evidently widely distributed during Pliocene times in Europe, India and China, and is also known from Pleistocene horizons in Europe (including England), North Africa and Java.

From these palæontological data, it is evident that the Cercopithecidæ had a wide distribution in Tertiary times. Especially noteworthy is the presence of baboons in North India in Pliocene times, and of *Macacus* in North Europe even so late as the Pleistocene age.

The Platyrrhine monkeys, which are to-day confined to South America, reaching as far north as Southern Mexico, are divided into two families, the Cebidæ and the Hapalidæ. The latter comprise the marmosets, of which several genera are recognized. They include the smallest of living monkeys, being about the size of squirrels, and are characterized by the presence of sharp curved claws on all the digits except the hallux (which bears a flattened nail), and by the loss of the third molar tooth. Evidence will be adduced which

* Formerly, however, referred to the Miocene.

indicates that the marmosets are in many respects the most primitive of living monkeys.

The Cebidæ are larger animals in which the third molar is retained as in all other Anthropoidea. They are thoroughly arboreal in habits, and many of them have developed prehensile tails by means of which they can hang and swing from the branches, or which they can even use as a " third hand " for grasping objects such as food. One genus, *Callimico*, resembles the marmosets in the possession of sharp claws, and, on the basis of this, Pocock[5] has included it in the Hapalidæ. But the dentition corresponds with that of the Cebidæ and thus affords a sharp contrast with the marmosets, probably outweighing the evidence of the external characters. A number of sub-families have been recognized among the Cebidæ, but it will suffice for the present purpose to note a few of the better known genera. *Cebus* includes the small capuchin monkeys which are familiar as pets. The squirrel monkeys (*Chrysothrix* or *Saimiris*) are small animals characterized among other features by an exceptional elongation of the skull. *Nyctipithecus* (or *Aotus*) represents a genus which is nocturnal in habits, and in association with this the eyes are rather large. These monkeys are colloquially known as Douroucoulis. The Teetee monkeys (*Callicebus* or *Callithrix*) are found on the banks of the Amazon and superficially are not unlike *Nyctipithecus* except that the eyes are smaller and the nostrils wider apart. *Pithecia* includes the Saki monkeys ; this genus is characterized by the length of the tail, which is not prehensile. The spider monkeys (*Ateles*) represent the culmination of the tendency towards the type of arboreal specialization which is manifested by the Platyrrhini generally. The tail provides a naked sensory surface on its under aspect near the tip and is marvellously prehensile, and the thumb is either completely absent or is extremely vestigial. Lastly, in the genus *Alouatta* (or *Mycetes*) is represented the Howler monkey, which is the largest of the Platyrrhines. The skull of this monkey is unusually modified in association with the development of an enormously inflated hyoid bone. The latter functions as a resonating chamber by means of which an exceptionally great volume of vocal sound can be emitted.

The geological record of the Platyrrhines is extremely scanty. Except for a portion of the skull and dentition of a Miocene form from Patagonia, *Homunculus*, which is very similar to the modern small Cebidæ, practically nothing is known directly of their early palæontological history.

The Tarsioidea

This sub-order is represented by one living genus, *Tarsius*, the little tarsier which is found in the Malay Archipelago. This animal—which is the size of a two-weeks-old kitten—has attracted much attention by reason of the fact that it combines a number of remarkably primitive with unexpectedly advanced characters. For many years it had been relegated to a family of the Lemuroidea, while some authorities (*e.g.*, Hubrecht, Pocock, Wood Jones) would associate it closely with the Anthropoidea. The consensus of opinion to-day recognizes that *Tarsius* occupies a systematic position somewhere between the lemurs and the monkeys, and this view is best expressed by accommodating it in a central sub-order of the Primates, Tarsioidea.[3]

The tarsier is a nocturnal and entirely arboreal creature.[2] It exhibits marked specialization in the enormous size of its eyes, and in the peculiar modification of its hind-limbs for leaping among the branches. The ears are large, and the tail (which is not prehensile in the usual sense of the term) is long and naked except for a terminal tuft of hair. In the structure of the nose and upper lip it resembles the Anthropoidea and contrasts rather strongly with the Lemuroidea (Fig. 4 and Fig. 58).

At the beginning of the Tertiary epoch, tarsioids were widely distributed over the world, and over twenty different genera have been recorded from Palæocene and Eocene deposits. Of these fossils rather more than half are derived from North America, and the remainder from Europe. Only one genus, *Omomys*, is common to both hemispheres. It is important to emphasize that some of the early tarsioid genera are only known from quite small fragments of the jaws with portions of the dentition. The tarsioid nature of these forms has been inferred from the character of the teeth, but, as Matthew has pointed out, " it is yet to be proved that they are really in a Primate stage of evolution in skull and skeleton or in brain characters." For the purpose of discussion we may accept the conclusions of the competent authorities who have studied them and assigned them to the Tarsioidea, with the reservation that a more extensive palæonto-

FIG. 4.—A YOUNG TARSIER.

From a photograph by the author.

logical record may call for a revision of the diagnosis in a few cases. In some other fossil tarsioids, such as *Necrolemur*, *Tetonius* and *Pseudoloris*, the skull is known in whole or in part, while in others (*e.g.*, *Omomys*) fragments of the limb skeleton have been found which add corroborative evidence to that of the dentition.

In the early tarsioids of which the skull and skeleton are at all preserved, the characteristic specializations of the skull and hind-limb of the modern tarsier are already in evidence. Perhaps the most interesting fossil representatives of the sub-order are certain European forms—*e.g.*, *Necrolemur* and *Microchœrus*, which are assigned to a separate family, Microchœridæ. In some structural features they were much more pithecoid than *Tarsius*, and the group as a whole may have formed the basis from which the Anthropoidea took their evolutionary origin. Most of the other fossil tarsioids are allocated to the Anaptomorphidæ, many of which exhibit divergent specializations of the teeth which may be rather extreme. Nothing is known of the palæontological record of the Tarsioidea after the Eocene.

The Lemuroidea

The members of this sub-order display their Primate status much less obtrusively than the monkeys, and in many respects they seem to represent a midway stage between monkeys and lower mammals. It is for this reason that German naturalists refer to them as " Halbaffen." In past years the question has been vigorously discussed whether lemurs should be included in a common group with the Anthropoidea, or whether, indeed, they should not properly be excluded from the Primates altogether and accommodated in a separate mammalian Order, the Lemures. Even to-day this latter view is maintained by at least one prominent zoologist, but the consensus of opinion is against it. In later chapters the evidence of the Primate status of the lemurs will be reviewed in detail, but we may note at this juncture that in their evolutionary history they have exhibited developmental tendencies which are remarkably similar to those of the Anthropoidea. In their external characters, the lemurs resemble the monkeys in the functional adaptation of the manus and pes for grasping purposes, in the freedom and mobility of the pollex and hallux, and in the presence of flattened nails on the digits. In all except one lemur (*Chiromys*), only the second digit of the foot is provided with a sharp claw.   On the other hand, a strong contrast to the monkeys (and a corresponding approximation to non-Primate mammals) is provided by the elongated snout which projects forwards well beyond the level of the chin, the naked and moist skin of the muzzle, the median cleft in the upper lip which is bound down to the

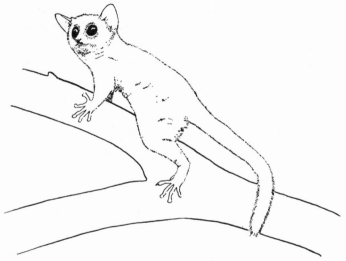

FIG. 5.—A MOUSE LEMUR (*Microcebus murinus*).
From a photograph by F. W. Bond.

underlying gum, the large and freely-moving ears, the position of the eyes, and the immobile expression of the face.

The lemurs are almost all entirely arboreal creatures.   They are also nocturnal in their activities, and this is evidenced by the rather large size of the eyes.

The recent lemurs can be divided into two main series, the Lemuriformes and Lorisiformes, which are distinguished by certain fundamental characters of the skull, nasal cavity, and external genitalia.   The Lemuriformes are confined today entirely to Madagascar and the neighbouring Comoro Islands, and are often referred to as the Malagasy lemurs. They comprise three families, the Lemuridæ, Indrisidæ and Chiromyidæ (or Daubentoniidæ). The Lemuridæ are again divided into two sub-families, the Lemurinæ and Cheirogaleinæ. The former contains the genus *Lemur*, which is not so completely arboreal or nocturnal in its habits as other genera. Nevertheless, it is a very active climber and is provided with a long tail which functions as a balancing mechanism. The sub-family Cheirogaleinæ includes the tiny mouse lemurs (*Microcebus* and *Cheirogaleus*), which probably represent the most generalized of the living lemuriforms (see Fig. 5).

The Indrisidæ differ from the Lemuridæ in their dental formula, and are more thoroughly arboreal. This family includes the largest of the modern lemurs.

The Chiromyidæ are represented by one species only, the Aye-aye (*Chiromys* or *Daubentonia*). This peculiar and aberrant lemur is characterized by the rodent-like structure of its front teeth (see Fig. 12), and by the retention of claws on all its digits with the exception of the hallux. In many points of its anatomy it departs widely from the other lemuriforms, and some authorities have even suggested that it represents a separate sub-order of the Primates—Chiromyoidea.[1] But its close affinity with the Lemuroidea, and especially with the lemuriform group, is betrayed by a number of highly significant characters in the skull, limbs and viscera.

The Lorisiformes are represented by two families, Lorisidæ and Galagidæ. The lorises are extremely slow-moving creatures, adopting a very deliberate crawling gait as they climb among the branches. The fore- and hind-limbs are of almost equal length, the snout is somewhat abbreviated, and the eyes are unusually large. The genus *Loris* is found in India and Ceylon, while *Nycticebus* extends from India over Malaysia. The Galagidæ (consisting of three genera) are confined to the tropical regions of Africa. *Galago*—an East African lemur—is characterized by the extreme modification of the hind-limb for jumping, and in this respect parallels very closely the tarsioid condition. *Perodicticus*, which includes the Pottos, is found in Sierra Leone and the Congo region, while a closely allied form, *Arctocebus*, inhabits West Africa.

Fossil lemuroids are known from early Tertiary deposits of Europe and America. The Lemuriformes are represented by a widespread family, the Adapidæ, which can be divided into two sub-families, the Adapinæ and the Notharctinæ. The former is confined to the Old World and is represented by two well-defined genera, *Adapis* and *Pronycticebus*. These have been found in Eocene deposits in France and are known by a number of well-preserved skulls with the dentition. Several species of *Adapis* have been described, of which the commoner are *Adapis magnus* and *Adapis parisiensis*. *Pronycticebus* is noteworthy because of its remarkably primitive features. From lower Eocene (or perhaps Palæocene) levels a fragmentary lower jaw and dentition of a primitive lemuroid, *Protoadapis*, have been described. It is generally agreed that this genus is really not closely related to *Adapis*, but probably has affinities with the American genus, *Pelycodus*.

The Notharctinæ comprise *Notharctus*, *Pelycodus*, and possibly *Aphanolemur*—all American fossils. *Aphanolemur* is known from the skull, but the teeth were not preserved in the specimen. *Pelycodus* is represented by several species, of which the earliest is referred to the basal Eocene or Palæocene and the others to the early Eocene. They were very primitive Primates, and the whole group is probably to be regarded as a direct precursor of the more advanced genus *Notharctus*. The fossil remains of the latter are almost complete and have been described in an illuminating monograph by W. K. Gregory. The Notharctinæ show rather strong structural contrasts with the Adapinæ and almost certainly represent an aberrant line of evolution.

During the Pleistocene age in Madagascar, a number of remarkable extinct forms existed. One of these, *Megaladapis* (belonging to the family Lemuridæ), reached a gigantic size.

Indeed, it is the largest Primate known and even exceeded the gorilla in bulk. *Palæopithecus* and *Mesopropithecus* are referable to the Indrisidæ, while a separate family, Archæolemuridæ, has been created to accommodate *Archæolemur* (= *Nesopithecus*) and *Hadropithecus*, which are also Pleistocene Malagasy lemurs of an aberrant type. Both *Mesopropithecus* and *Archæolemur* have attracted attention because of the monkey-like appearance of their skull characters.

During the Palæocene and Eocene ages of Europe and North America, there flourished a most interesting family, the Plesiadapidæ. The rather scanty remains of at least sixteen different genera have been recorded. By the late Dr. W. D. Matthew these forms were regarded as extinct representatives of the modern tree-shrews. Most authorities now assign them definitely to the Primates,[7] and some would group them together with the Aye-aye in the sub-order Chiromyoidea. For one of the characteristic features which they exhibit is the tendency to develop specialized front teeth which bear a striking resemblance to those of the Aye-aye. Indeed, there is considerable reason to infer that some of the plesiadapids may be closely related to the direct ancestors of this aberrant lemur.

Before concluding this brief outline of the Primates, mention may be made of the tree-shrews, because they will be referred to in later chapters quite frequently. These small and generalized arboreal creatures are found in Southern Asia. Although they are commonly included in a sub-order (Tupaioidea) of the Insectivora, there is abundant evidence to suggest that they are really very primitive members of the Order Primates. This evidence will be reviewed in Chapter X.

References

1. ABEL, O. : Die Stellung des Menschen im Rahmen der Wirbelthiere. Jena, 1931.
2. CLARK, W. E. LE GROS : Notes on the Living Tarsier. Proc. Zool. Soc., 1924.
3. ELLIOT SMITH, G. : The Zoological Position and Affinities of Tarsius. Proc. Zool. Soc., 1920.
4. MATTHEW, W. D. : The Evolution of the Mammals in the Eocene. Proc. Zool. Soc., 1928.
5. POCOCK, R. I. : On the External Characters of the South American Monkeys. Proc. Zool. Soc., 1926.
6. SIMPSON, G. G. : A Catalogue of the Mesozoic Mammalia.   London, 1928.
7. SIMPSON, G. G. : A New Classification of Mammals.   Amer. Mus. of Nat. Hist., vol. lix., 1931.
8. WEBER, M. : Die Säugethiere. Jena, 1928.
9. WOOD JONES, F. : Man's Place among the Mammals. London, 1929.

CHAPTER III

THE EVIDENCE OF THE SKULL

THE distinguishing features of the Primate skull are almost all related to the fact that in this group of animals there is a progressive tendency towards an enlargement of the brain, a high development of the mechanism of vision with a corresponding reduction in the apparatus of smell, the adoption to various degrees of an upright position of the trunk in locomotion, and the use of the fore-limb for prehension rather than for progression. There is evidence, moreover, that all these characters are directly associated with an arboreal mode of life, as Elliot Smith has demonstrated so clearly.[3]

The brain-containing part of the skull—the neurocranium—is usually rather voluminous and rounded, and provides sufficient surface for the attachment of the masticatory muscles without the development of bony crests and flanges such as are found in many mammalian skulls. Moreover, the masticatory muscles themselves are usually relatively small in correlation with a reduction in the size of the jaws. This reduction is shown in a recession of the snout region of the skull, which is no doubt partly determined by the diminishing importance of the nasal cavities, and also by the fact that in the Primates the hands are employed for much of the work which—in other mammalian Orders—is usually carried out by the teeth and lips. Thus the tooth-bearing part of the skull is on the whole of moderate size only.

The orbit is rather large, is surrounded (in all recent forms) by a bony ring formed by the articulation of the frontal with the malar bone, and the orbital aperture usually looks somewhat forwards. The lachrymal bone shows a tendency to reduction in size and to be withdrawn from the facial surface of the skull into the orbital cavity. The zygomatic arch is relatively slender.

The foramen magnum on the base of the skull, instead of facing almost directly backwards as is the case in completely pronograde mammals, becomes displaced forwards so as to

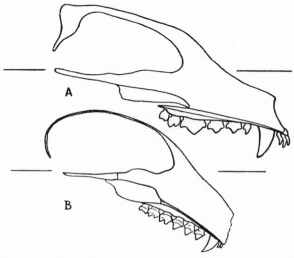

FIG. 6.—DIAGRAMMATIC LONGITUDINAL SECTION THROUGH THE SKULL OF A, A DOG, AND B, A LEMUR, TO SHOW HOW, IN THE LATTER, THE FACIAL REGION IS BENT DOWNWARDS ON THE BASICRANIAL AXIS.

This bending downwards of the face is more marked in the higher Primates, and is very characteristic of the Order as a whole.

look more downwards. This displacement of the foramen magnum is in part associated with a change in the poise of the head, and, related to it, is a bending of the basi-cranial axis so that the face—in addition to a general reduction in size—comes to be placed rather below the neurocranium than directly in front of it. This recession of the facial part of the skull is illustrated diagrammatically in Fig. 6, which shows a sagittal section through the skull of a lemur compared with that of a dog. But the relegation of the foramen magnum from its posterior position to the basal aspect of the cranium (which is also indicated in Fig. 6) is partly conditioned by or associated with the progressive development of the occipital lobes of the cerebral hemispheres which results from the expansion of the visual cortex. Thus, in most Primates, the occipital region of the skull is rounded and prominent, projecting well back behind the level of the foramen magnum.

Lastly, it may be noted here that the tympanic cavity (which contains the ossicles of the middle ear) is completely encased in bone, the osseous floor of the cavity being formed by an expansion of the petrous bone.

In considering these generalizations on skull structure, it is important to bear in mind that, within the limits of the Primates, there is quite a considerable variation in the details and proportions of the cranium, especially if fossil members of this Order are taken into account. We may best deal with these variations by discussing the skull in the different groups of Primates separately.

Lemuroidea

In general, the most primitive expression of the Primate skull is to be found in the lemuroids. In this group, as in lower mammals, the facial part of the skull is relatively large in comparison with the neurocranium. These animals thus often have a conspicuous and elongated snout, which is best shown in the Lemuriformes. In the lorises and galagoes (especially in the little *Galago elephantulus*), however, the snout has been reduced to a degree which parallels the condition seen in higher Primates—*e.g.*, the tarsier. In all the living lemurs, with the exception of *Chiromys*, the upper incisor teeth have undergone marked degeneration, and, with this, the premaxillary bones have become reduced to very small

proportions.*   This atrophy has not progressed so far in the
fossil Adapidæ and is not present at all in *Chiromys*.

The orbital aperture is large, in association with nocturnal
habits, and is directed somewhat forwards.   In the lorises it
looks upwards to a marked degree, a position which is also
to be seen in the fossil *Adapis parisiensis* (see Figs. 7 and 11).
The lachrymal bone has usually a moderate facial extension
in living lemurs and is said to be absent altogether in *Loris*

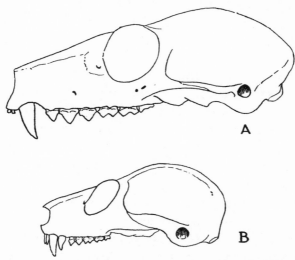

FIG. 7.—LATERAL VIEW OF THE SKULL OF A, *Lemur varius* × ¾, AND
B, *Perodicticus* × ¾.

Note the relatively elongated snout, the size and direction of the orbital
aperture, the rounded shape of the occipital region, and the reduction of the premaxillary region.

and *Nycticebus*.   It articulates directly with the malar bone
in the Lemuriformes, but in the Lorisiformes, as in higher
Primates, may be (at least in the adult) separated from it by
the maxilla, which thus takes a part in the formation of the
orbital margin.   The lachrymal foramen in all these recent
forms is situated without the margin of the orbit.

* In *Microcebus* and *Chirogale*, the degeneration of the incisor teeth
is less extreme, and in these forms the premaxilla is correspondingly less
reduced.

The sutural pattern in the orbito-temporal region of the skull is important to note, for it offers valuable evidence regarding affinities.   In primitive mammals (e.g., lipotyphlous insectivores and rodents) the inner wall of the orbit is formed mainly by the frontal and maxillary bones, which are in contact with each other by a long sutural junction, while anteriorly the lachrymal and posteriorly the orbito-sphenoid, alisphenoid and palatine bones also take a part in its formation (see Fig. 8, A).   Now, in animals in which the eyes are large and become to some extent rotated forwards (e.g., Carnivora and Primates), the maxilla comes to form the floor rather than the medial wall of the orbit and is thus separated from the frontal by a wide interval.   This interval may become filled up in two different ways.   In one case, the orbital plate of the palate bone extends forwards so as to reach the lachrymal bone (Carnivora, tree-shrews, Lemuriformes), and, in the other, the ethmoid bone becomes exposed in the orbit, separating the palatine from the lachrymal (lorises and galagoes, Tarsioidea and Anthropoidea).   Thus, by reference to this feature, the Lemuriformes (with the exception of Chiromys and Microcebus) are rather sharply separated from the Lorisiformes (see Fig. 8, B and C).   It may be conjectured that the divergent lemuriform and lorisiform types could only have been derived from a primitive and common type in which the fronto-maxillary contact in the orbital wall was as yet undisturbed, and there is interesting corroborative evidence to support this conception.   For in the Eocene Adapinæ (but not, it seems, in the Notharctinæ), the primitive arrangement of sutures in this region of the skull may be found.   In a specimen of Adapis parisiensis in the British Museum (No. M 1345) and in Pronycticebus, for instance, it is quite apparent that the orbital plate of the palatine is small, and a fronto-maxillary contact separates it widely from the lachrymal.   Further, this is also the condition in the orbit of the modern Chiromys,*

* It may thus be argued that the Chiromys stock must have separated from the main lemuriform stock at least at the Adapis stage of evolution.

which thus does not conform to the typical lemuriform arrangement. Kollmann[8] has recorded a somewhat intermediate state of affairs in the mouse-lemur (*Microcebus*) in which both the os planum of the ethmoid and the palatine bone intervene between the frontal and maxilla, the palatine not quite reaching as far forwards as the lachrymal.

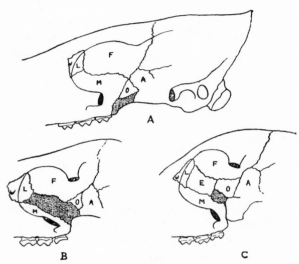

FIG. 8.—DIAGRAMS SHOWING THE BONY ELEMENTS WHICH MAY CONTRIBUTE TO THE FORMATION OF THE ORBITO-TEMPORAL REGION OF THE SKULL.

The palatine bone is indicated by stippling: A, Primitive mammalian condition, in which there is a wide fronto-maxillary contact. B, The condition in recent Lemuriformes, in which the orbital plate of the palatine articulates with the lachrymal. C, The condition in Lorisiformes, *Tarsius* and the Anthropoidea, in which the ethmoid separates the frontal bone from the maxilla, and the palatine from the lachrymal.

A, Alisphenoid; F, Frontal; M, Maxilla; L, Lachrymal; O, Orbitosphenoid.

On the side wall of the skull in all lemurs the alisphenoid extends up to articulate with the parietal bone, and thus the frontal and temporal bones do not come into contact—a primitive and generalized mammalian feature.

The structure of the auditory region of the mammalian

skull is of the highest importance in the working out of systematic affinities—as was shown so conclusively in the monumental work of van Kampen.[7]   In a primitive mammal, (*e.g.*, certain Insectivora and Edentates) the tympanic cavity, which is hollowed out from the petrous bone, has no ossified floor, while its lateral wall is formed by the tympanic membrane or ear-rdum which is supported in a fine bony ring, the os ectotympanicum (Fig. 9, *a*).   In most mammals, however, the membranous floor of the tympanic cavity becomes ossified, and this ossification may involve different bony elements of

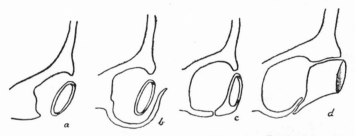

FIG. 9.—SCHEMATA ILLUSTRATING THE VARIOUS RELATIONS OF THE TYMPANIC RING TO THE TYMPANIC BULLA.

(*a*) Primitive mammalian condition, in which the ring is exposed and the floor of the tympanic cavity is unossified. (*b*) The lemuriform type, in which the ring is enclosed within the osseous bulla. (*c*) The lorisiform and platyrrhine type, in which the ring remains exposed and contributes to the formation of the outer wall of the bulla. (*d*) The tarsioid and catarrhine type in which the ring is produced laterally to form a tubular auditory meatus.

the skull in different groups.   In the Primates the osseous floor is formed mainly from the petrous bone itself, and in the lower members of the Order this becomes distended to form a smooth rounded swelling on the base of the skull, which is called the tympanic bulla (see Figs. 9 and 13).

In the Lemuriformes the bulla is relatively large and, indeed, grows over the tympanic ring so that the latter becomes completely enclosed by it (Fig. 9, *b*).   In the Lorisiformes, on the contrary, the tympanic ring retains its primitive position on the surface and, becoming somewhat expanded and applied to the outer margin of the tympanic expansion of the petrosal, contributes to the formation of the bulla wall (Fig. 9, *c*). In *Loris* and *Nycticebus*, the ectotympanic is further produced to form a very short tubular auditory meatus. Thus, again, the Lemuriformes and Lorisiformes stand in contrast with each other. This contrast also involves the arrangement of certain foramina on the base of the skull, which we may now consider. In close relation to the floor of the tympanic cavity is an artery, the entocarotid (which corresponds to the internal carotid artery of human anatomy). This artery gives off a branch, the stapedial artery, which runs up through the stapes to reach the roof of the tympanic cavity, while the main artery itself continues on as the arteria promontorii, eventually to reach the intracranial cavity and contribute to the blood supply of the brain. The mode of formation of the bulla determines to some extent the manner in which the entocarotid artery and its branches enter the skull (as explained by Winge). In the Lemuriformes, the entocarotid artery passes into the tympanic cavity by a foramen situated at the posterior margin of the bulla close to the stylomastoid foramen (see Fig 13, A). Inside the bulla it divides to form a large stapedial artery and a very small arteria promontorii. This marked atrophy of the latter artery is a lemuriform specialization of high significance, and is associated with the fact that in this group the blood supply of the brain is derived almost entirely from the vertebral arteries. In the Lorisiformes, the entocarotid artery divides before it reaches the base of the skull to form a large arteria promontorii and a very small stapedial artery—a proportionate division which is precisely the reverse of the condition in the Lemuriformes. Moreover, in the lorisiform type, the arteria promontorii does not run a tortuous course through the tympanic cavity (as it does in lemuriform and primitive mammals) but takes a short cut by passing directly into the skull through the middle lacerated foramen. This foramen, it should be noted, is not present in the Lemuriformes, for in them it is covered over by the base of the bulla. In the Lorisiformes, the small stapedial artery pierces a minute foramen on the posterior aspect of the bulla and runs the same course as in the Lemuriformes.

It is interesting to record that, among the Lemuriformes, the Cheirogaleinæ show an arrangement of these arteries which appears to correspond to the lorisiform condition, though in other features the tympanic region of this sub-family is essentially of the lemuriform type.

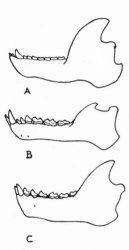

A

B

C

Fig. 10.—The Mandible of A, a Lemuroid (*Adapis parisiensis*) × ½, B, *Tarsius* × 1¼, and C, a Fossil Monkey (*Parapithecus*) × ¾.

The mandible in the Lemuroidea shows characteristically a rather slender horizontal ramus and a broad but low ascending ramus with well - marked coronoid and angular processes. The horizontal rami do not diverge from the symphysis to more than a slight degree. The symphysis menti is unossified in all modern lemuroids, but in the extinct Archæolemuridæ and Adapinæ it is stated to be synostosed in the adult individual. The significance of this bony fusion is not clear, for there is no evidence that it is directly related to the nature of the dentition or to the size of the mandible.

Attention may now be turned to the skull form of primitive fossil lemuroids, some details of which have already been referred to above.

In the extinct family Adapidæ, the skull structure of *Adapis* and *Notharctus* (both of Eocene age) is known in considerable detail. In both genera the structure of the tympanic region is precisely as it is in the Lemuriformes, and there can be no doubt that they should be referred to this group. The orbits are smaller than in modern lemurs, and this may be taken to indicate that nocturnal habits had not been fully adopted by them. The brain-case is also considerably smaller, and, coincidently with this, conspicuous sagittal and lambdoidal crests are produced in order to provide accommodation for the attachments of the temporal and nuchal muscles. The postorbital constriction of the skull, as viewed from above, and the small size of the frontal bones, are very striking outward manifestations of the relatively poor cerebral development.* Reference has already been made to the more primitive condition of the premaxilla in the Adapidæ, and of the orbito-temporal region in *Adapis* and *Pronycticebus* (*vide supra*, p. 44). The lachrymal bone in both *Notharctus* and *Adapis* is markedly reduced and is limited to the orbital cavity, while the lachrymal foramen is at or immediately within the orbital margin. This reduction of the lachrymal is generally regarded as an advanced character which it is curious to find in Eocene lemurs, and has led some authorities to doubt whether the Adapidæ could possibly be ancestral to modern lemurs.

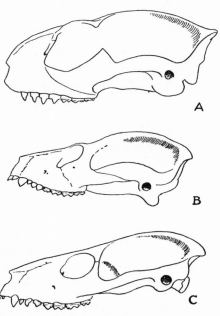

Fig. 11.—Lateral View of the Skull of Fossil Adapidæ.

A, *Pronycticebus gaudryi* × 1 ; B, *Adapis parisiensis* × ⅔ ; C, *Notharctus osborni* × ⅔ (after W. K. Gregory). Note the small brain-case, the muscular ridges, the unreduced premaxillary region, and in B and C the small size of the orbits.

Gregory, indeed, has argued

* In the fragmentary skull of the American Eocene genus *Aphanolemur*, the brain-case is relatively more expanded and the sagittal crest has partly disappeared as the result of the separation anteriorly of the temporal ridges. But the status of this fossil is uncertain, for the dentition was not preserved.

that the greater facial extension of the lachrymal in modern lemurs is a secondary condition which is associated with the increase in the size of the eye and the consequent displacement on to the face of the lachrymal sac and duct. But this must be considered an open question, for it is certainly not supported by the evidence of *Pronycticebus*, which is also an Adapid and considerably more primitive than either *Adapis* or *Notharctus*.

The foramen magnum in *Notharctus* looks backwards, instead of downwards and backwards as it does in modern lemuroids, and in this feature it resembles the condition in completely pronograde mammals. In other words, there is evidence to show that in Eocene times the characteristic poise of the head which becomes progressively developed in the Primates had not yet been acquired to even a slight degree.

It will be observed from Fig. 11 that on the whole the skull structure of *Adapis* and *Notharctus* is much more primitive than in modern lemuroids and takes us back some considerable way towards the ancestral form from which the latter have diverged in the course of their evolutionary development. This is still more the case with the interesting skull of *Pronycticebus* which was found in Eocene deposits in France. This genus was described in 1904 by G. Grandidier, who was disposed to believe that it might represent a precursor of the modern *Nycticebus*. In 1916 Stehlin, on the basis of tooth structure, related *Pronycticebus* to *Anchomomys* (a fossil representative of the Tarsioidea), and, as his opinion was accepted by several authorities, the fossil has become generally to be regarded as a remarkably primitive tarsioid. My own studies on the skull of *Pronycticebus* (for the opportunity of which I have to thank Professor Boule of the Natural History Museum of Paris) indicate quite clearly that it is a true lemuriform and that it really represents a very primitive member of the Adapinæ.[1] The neurocranium is rather broad (approximating in this respect to the tarsioids), but poorly developed in relation to the skull as a whole, the orbits are of moderate size and proportionately larger than in other Adapids, the foramen magnum occupies a position intermediate between that of *Lemur* and *Notharctus*, the postorbital constriction is marked, and the orbit is not encircled by a complete bony ring (Fig. 11, A). The dentition is of a very generalized character. The great interest of *Pronycticebus* lies in the fact that it is a remarkably primitive Primate skull, but yet (on the basis of the structure of its tympanic region) definitely assignable to the lemuriform group of the Lemuroidea. It thus provides strong evidence suggesting that the Lemuriformes had already separated from the lorisiform stock in a process of divergent evolution at a stage when the brain-case was comparatively small, the dentition of a very generalized character, and the postorbital bar of bone as yet incomplete.

The fragmentary jaw remains of the early Notharctinæ represented by the genus *Pelycodus* (reaching back to basal Eocene antiquity) show that in many ways this form was still more primitive. But perhaps the most suggestive palæontological evidence which indicates the separation of the Lemuroidea from a common Primate ancestry and their differentiation into divergent groups at an extraordinarily early stage in the history of the Order is supplied by the remains of the Plesiadapids. It has already been recorded that the Plesiadapidæ are now commonly included among the Primates by systematists, and that by some palæontologists they are grouped together with *Chiromys* to constitute a separate sub-order, the Chiromyoidea. The evidence for the latter association (apart from the creation of a separate sub-order) seems to be reasonably sound, for the Plesiadapidæ provide a rather remarkable illustration of some of the stages through which the ancestral forms of *Chiromys* must have passed in the progressive specialization of its dentition and are, moreover, found in geological deposits of an age when such changes might be expected to have occurred. The mandible and front portion of the skull of one member of this family,

Stehlinella (= *Stehlinius*), were found in 1921 in the upper Eocene deposits of Utah.   Fig. 12, A represents this skull as it presumably must have appeared when it was complete. It will be noted that, as in *Chiromys* (Fig. 12, B), the premaxilla is of large size, and reaches up almost to the orbital margin.

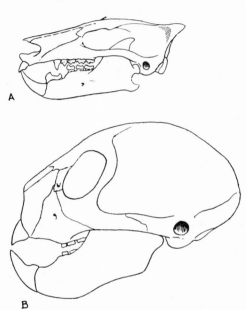

A

B

FIG. 12.—LATERAL VIEW OF THE SKULL OF A, *Stehlinella* (partly reconstructed from the figure by W. D. Matthew) × 1 ; B, *Chiromys* × ⅔.

Note the highly specialized front teeth, the reduction of the back teeth, and the large size of the premaxilla in both skulls.   Note also the remarkably primitive form of the plesiadapid skull.

On the other hand, the skull is remarkably primitive as a whole, closely resembling in general appearance and proportions the skulls of primitive insectivores such as *Gymnura*. The orbit is not surrounded by a bony ring, the braincase is very small, while the facial part of the skull is large and projecting.

If *Stehlinella* is accepted as an early representative of the *Chiromys* stock, and if the *Chiromys* stock forms a natural subdivision of the Lemuriformes (as is indicated by the construction of the tympanic region of its skull), then it can only be inferred that the Lemuriformes and the Lorisiformes commenced their divergent specialization when the skull had certainly not progressed beyond the primitive stage exemplified by *Stehlinella*.

Let us now review the general inferences which may be drawn from the lemuroid skull in regard to the cranial characters of the ancestral forerunner of the Lemuroidea and in regard to the splitting up during the course of evolution of this group into natural subdivisions. In doing so, we may with propriety assume that evolution has on the whole been marked by a regular and orderly sequence of morphological changes from primitive to specialized features, and that it is reasonable to suppose that this evolutionary progress has followed definite lines which are to a large extent limited by the law of irreversibility. There can be little doubt that the common lemuroid ancestor was a small-brained animal, and that the neurocranium was correspondingly limited in volume, with a well-marked postorbital constriction. Presumably, therefore, the skull was characterized by well-developed sagittal and lambdoidal crests in order to provide sufficient surface for the attachment of masticatory and nuchal muscles.* The premaxillary element was at least as large as it is in the Adapidæ, subsequently undergoing a reduction in all the lemuroids except for the Chiromyidæ (and Plesiadapidæ) in which it became hypertrophied. The skulls of *Pronycticebus* and *Stehlinella* suggest that the orbit was not yet surrounded by a bony ring and was not unusually large, while the orbital aperture was directed laterally. The orbitotemporal region of *Adapis* and *Pronycticebus* shows that the articulations of the bony elements in this region were of a primitive nature (which, indeed, would also be anticipated from a consideration of the divergent types of sutural pattern in the Lemuriformes and Lorisiformes). The foramen magnum occupied a primitive position, looking directly backwards as in pronograde mammals. The construction of the tympanic region in modern and fossil lemurs provides a further indication of the primitive nature of the skull in this hypothetical common ancestor. It has been pointed out that primitively

---

* In the Pleistocene genus *Megaladapis* these crests are conspicuously developed. In this case, however, the feature is no doubt of a secondary nature, in correlation with the large size of the animal (*vide infra*, p. 61).

the ectotympanic bone lies on the surface, but that in the Lemuriformes it becomes enclosed by the tympanic bulla which grows over it.   This feature—which is found elsewhere among the Mammalia only in the tree-shrews—must be regarded as a peculiar and far-reaching structural specialization, and it is difficult to suppose that from such a specialized arrangement a more generalized condition in which the ectotympanic is outside the bulla could again have been derived.   Thus it seems impossible to imagine that the Lemuriformes represent in any way an ancestral stock which might have given rise to other Primates.   Gregory, however, has argued that the higher Primates might have been derived from a form such as *Notharctus* by imagining that, with the increase in breadth of the skull base, the bulla became displaced medially towards the mid-line, while the ectotympanic element—in virtue of its attachment to the squamous portion of the temporal—became drawn out from the bulla and thus once more set free.   Such a process appears unlikely, for it may be observed that in some of the large skulls of the Pleistocene lemurs of Madagascar the base of the skull underwent a considerable increase in width, but, in spite of this, there is no evidence that the extrication of the tympanic ring has occurred in this instance.*   It is highly improbable, in fact, that in the course of evolutionary change the tympanic element would become disengaged from the bulla once it had become caught up inside it, and it may be inferred, therefore, that as far as the evolutionary origin of the higher Primates is concerned, the Lemuriformes can hardly come in for consideration.   It is also not fully realized, I think, that there are other lemuriform specializations intimately associated with these modifications in the tympanic region.   Thus the atrophy of the arteria promontorii is a strikingly divergent

* In *Palæopropithecus*, however, the tympanic ring is said to be incompletely enclosed by the bulla, but in this case the bulla is shrunken and deflated.   In *Megaladapis* the bulla is produced into a long meatus and, according to Stehlin's account, the ectotympanic remains inside, fused with the bulla wall.

specialization which would seem also to debar this group from a place in anthropoid (or lorisiform) ancestry, for in the higher Primates the progressive enlargement of this artery which comes to supply a large and important part of the brain is a remarkable feature.* The tympanic region in the Lorisiformes is certainly more primitive in so far as the ectotympanic bone remains superficial, and it is perhaps possible that from such a condition the lemuriform arrangement might have been derived. But the lorisiform ectotympanic is usually somewhat expanded to take part in the formation of the bulla, which in the Lemuriformes is constructed entirely from the petrosal bone. Moreover, the course of the entocarotid in relation to the bulla is fundamentally different in the two groups, and the direct passage of this artery through the middle lacerated foramen is in itself a divergent lorisiform specialization of considerable significance. Thus it would appear most reasonable to suppose—as suggested by Stehlin—that the divergent trends in the development of the tympanic region in Lemuriformes and Lorisiformes indicate a derivation from a common ancestral form in which the osseous bulla was still unformed—that is to say, in which the floor of the tympanic cavity was still membranous, as it is in some generalized and primitive living mammals. The conclusion from these considerations that the lemuriform and lorisiform stocks could neither have been ancestral to the other is fully borne out by the details of the orbito-temporal region, as has already been pointed out.

The nature of the evidence as a whole, then, suggests that the skull of an ancestral form which might have given rise to the different groups of the Lemuroidea was of a very primitive kind, resembling that of the most primitive insectivores

* Thus the enclosure of the ectotympanic within the bulla is but one detail in a whole complex of structural modifications. It is the *sum* of these modifications which makes it difficult to conceive that the lemuriform type of skull could have provided a basis for the evolutionary development of other types of skull in the Primates.

which are known.   Such a skull, indeed, would be indistinguishable from an insectivore skull except, possibly, by reference to dental characters.

Tarsioidea

The skull of the only living representative of this sub-order —*Tarsius spectrum*—is characterized by its enormous orbital cavities, the openings of which are directed forwards and outwards (Fig. 14, C).   It has been conjectured that many of the cranial characters in *Tarsius* which resemble those found in the higher Primates are really secondary to this aberrant feature.   Thus the enlargement of the eyes is associated with a compression of the nasal region so that the olfactory cavities are much reduced ; the neurocranium has been compressed in an antero-posterior direction so that, viewed from above, it is broad and rounded ; and this compression has apparently led to a flattening of the alisphenoid, which thus appears to take a much more obtrusive part in the formation of the posterior wall of the orbit than in the lemurs.   This disturbing influence of the large orbits makes it extremely difficult to determine whether some of the pithecoid resemblances of the tarsioid skull are, as it were, merely fortuitous, or whether they do in fact indicate a real structural affinity with the monkeys.

The reduction of the snout is at least partly illusory, for the posterior part of this region is overlapped by the orbits. The latter are not only surrounded by a bony ring, but are also partly shut off from the temporal region of the skull by an incomplete bony wall formed from expansions of the frontal, malar, and alisphenoid bones, approaching in this respect the Anthropoidea.   As in the Lorisiformes and *Microcebus*, the os planum of the ethmoid enters into the formation of the inner wall of the orbit, and the lachrymal is separated from the malar bone by the maxilla.   In the temporal fossa the alisphenoid articulates with the parietal bone.   The lachrymal has a slight facial extension, and the lachrymal foramen is outside the orbit.   The foramen magnum is displaced on to the basal surface of the skull and looks almost directly downwards, while the occiput is prominent and well rounded (see Fig. 14).   The bending of the basicranial axis, further, is much more marked than it is in lemurs.

The tympanic bullæ are relatively enormous, and extend in an antero-medial direction so that they almost meet in the mid-line and approximate to the palatal region (see Fig. 13, B).

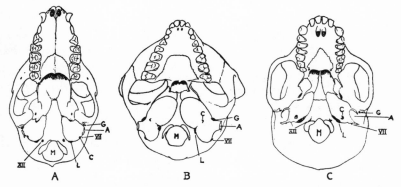

FIG. 13.—THE BASE OF THE SKULL IN A, *Lemur catta* × ½; B, *Tarsius* × 1; C, *Cebus fatuellus* × ½.

A, Auditory aperture; C, Entocarotid foramen; L, Posterior lacerated foramen; M, Foramen magnum; VII, Stylomastoid foramen; XII, Hypoglossal foramen.

The bulla is formed from the petrous bone, while the ectotympanic is applied to its outer surface superficially.   Moreover, the latter element is produced to form a very definite tubular meatus precisely as in the Old World monkeys (Fig. 9, D). The entocarotid artery enters the tympanic cavity by piercing the under surface of the bulla near its centre, and its course through the cavity corresponds to the condition found in the Lemuriformes, the arteria promontorii, however, being relatively large.

The mandible in the tarsioids is slightly built, with a relatively low and broad ascending ramus and a small coronoid process (Fig. 10, B). The horizontal rami diverge conspicuously from the region of the symphysis in correlation with the relatively broad cranial base which determines the wide separation of the glenoid cavities articulating with the mandibular condyles. In the modern *Tarsius* the symphysis menti is normally unossified (although the contrary is affirmed by Duckworth[2] and Wood Jones[11]), but in the middle Eocene genus *Cænopithecus* it was synostosed, in this way resembling the normal condition in the Anthropoidea.

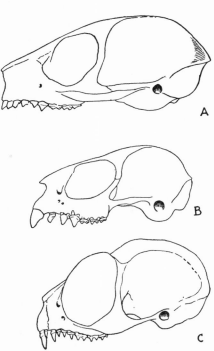

FIG. 14.—LATERAL VIEW OF THE SKULL OF A, *Necrolemur* (after Stehlin) × $\frac{4}{3}$; B, *Tetonius* (after Matthew, partly restored) × $\frac{4}{3}$; C, *Tarsius* × $\frac{4}{3}$.

Note the more primitive features in the fossil tarsioids as shown by the smaller reduction of the snout region, the smaller orbits, and the less rounded contour of the neurocranium.

The fossil specimens of certain tarsioid skulls provide us with an indication of the line followed in the evolutionary differentiation of the modern tarsier, and also emphasize the great antiquity of the specializations peculiar to this group. Thus the fragmentary remains of *Pseudoloris* from the upper Eocene demonstrate that at this early period the large orbits and shortened muzzle characteristic of the living *Tarsius* had already been acquired. The maxillary fragment of *Anchomomys* (middle Eocene) indicates an orbit of similar dimensions. In the Eocene *Tetonius* the orbit (as compared with *Tarsius*) is not so large, and it opens widely into the temporal fossa behind. The facial part of the skull is also less reduced, the neurocranium smaller and more constricted anteriorly, and the cranial roof flatter (see Fig. 14, B). In *Necrolemur* (also an Eocene tarsioid) a still more primitive condition is found in many respects (Fig. 14, A). The orbits are yet further reduced and the snout is longer. The brain-case appears more elongated as viewed from the side, while its breadth in a dorsal view is partly exaggerated by the inflation of the mastoid air cells (as in *Galago*). The cerebellar fossa is not so completely overlapped by the cerebral fossa of the skull, the foramen magnum looks downwards and backwards instead of almost directly downwards as it does in *Tarsius*, and the basicranial axis is more straightened out. There is a slight indication posteriorly of a sagittal crest. The tympanic bulla is not quite so large as in *Tarsius*, but, in curious contrast to the latter, the ectotympanic auditory meatus is longer and the entocarotid foramen is situated posteriorly in the bulla, occupying a position which is also found in the marmosets. These progressive characters in which *Necrolemur* approaches the monkeys more closely than does *Tarsius* are accompanied by a remarkably primitive feature, for, as Stehlin[10] has pointed out, the wall of the bulla is partly formed by a tympanic process of the alisphenoid, as in quite primitive lower mammalian skulls.

It has been noted in the previous section that *Pronycticebus*, of the upper Eocene, has been regarded as a member of the Tarsioidea on the basis of its dental characters. Although this opinion is no longer tenable, the fossil does represent an extraordinarily generalized type of Primate skull (with the exception only of the tympanic region), and there can be little doubt that the skull of the early ancestral tarsioids must have been not very dissimilar to it. In its general proportions, therefore, it provides us with a basis from which we may reconstruct mentally the steps along which the specialized skull of the modern tarsier has been developed.

The few fossil tarsioid skulls which are available for study certainly illustrate a progressive differentiation from a more primitive type to the remarkably pithecoid skull of *Tarsius*. Now, it is known from skeletal remains that, at least at the *Necrolemur* stage, the tarsioids had already started on a developmental trend towards the specialized locomotive apparatus associated with a saltatory mode of progression which quite definitely could not have formed a stage in the evolution of higher Primates (*vide infra*, p. 119). The suspicion is aroused, therefore, that many of the pithecoid traits of the skull of *Tarsius* may have arisen independently of the evolution of the monkeys. Indeed, it has already been suggested that not improbably some (perhaps most) of these " advanced " characters (such as the broad and rounded brain-case, the formation of a bony posterior wall to the orbit, the recession of the snout and the position of the foramen magnum) may be directly conditioned by the unusual development of the orbits, which has apparently led to a considerable distortion of other parts of the skull. On the other hand, the numerous similar tendencies in evolutionary development which the tarsioids show with the Anthropoidea (even though superimposed upon a foundation of cranial specialization peculiar to the tarsioids) would seem to indicate a derivation from an ancestral form with similar potentialities, and it may be legitimately inferred, therefore, that the modern *Tarsius* has been ultimately derived from a generalized precursor which also gave rise to the higher Primates. The immediate precursor of the tarsioid stock, at a point in evolution when the latter had first definitely separated off from the other prosimian stocks, may quite probably have possessed a skull structure not unlike the lorisiform type in its *general* plan and in the construction of the tympanic bulla (but, of course, lacking the characteristic reduction of the premaxillary region of the Lemuroidea). But the direct course of the entocarotid artery in the Lorisiformes—passing through the foramen lacerum medium without traversing the tympanic cavity—

is a specialized feature from which the more primitive condition in the tarsioid skull could hardly have been derived. It has already been pointed out, moreover, that the lemuriform type of skull is far too specialized to have formed a basis for the evolutionary development of other Primate stocks. Thus, on the evidence at hand, it is clear that the skull of the ancestral tarsioids could hardly be referred either to the lemuriform or the lorisiform groups of the Lemuroidea (at least as these groups are defined on the basis of skull structure). If such a skull were found without evidence of the nature of its dentition, it would presumably be regarded either as a primitive tarsioid, or as the skull of an insectivore with an unusually well-developed brain.

Anthropoidea

The simplest expression of the skull type of the Anthropoidea is to be found in some of the small New World monkeys (*e.g.*, *Hapale*, *Cebus*). The brain-case is proportionately large and, compared with the condition in lower Primates, dominates the relatively small facial part of the skull very strikingly (Fig. 15). There is an absence of bony crests, which, however, may be developed quite powerfully in certain of the larger members of this sub-order (*e.g.*, the gorilla). This formation of bony crests clearly depends on the relation between the bulk of the masticatory and nuchal musculature and the surface area of the neurocranium. Generally speaking, they tend to be developed in the smaller Primate skulls in which the brain is relatively small (as we have seen in previous sections) and in large Primate skulls in which the musculature is massive. They are thus by themselves of somewhat secondary importance in estimating degrees of affinity.

The frontal bone in the Anthropoidea forms a proportionately large part of the cranial roof in association with the increasing development of the frontal lobes of the brain. On the lateral wall of the cranium the sutural pattern varies.

In general, the alisphenoid articulates with the parietal bone as in primitive skulls. In some monkeys (but by no means all), and in the gorilla and the chimpanzee (but not in Man, the orang or gibbon), these bones are kept apart by a fronto-temporal articulation. It has been assumed by some authorities that the latter is a typical pithecoid feature, and it has thus been contrasted with the more primitive sutural pattern in the human skull. Such an assumption, however, is not correct, and, indeed, there may be considerable variation in a single species. Thus a fronto - temporal contact has been described as an abnormality in the gibbon skull, and a few cases of the human and primitive disposition of the sutures in this region with parietoalisphenoid contact has been recorded in the gorilla. It may be legitimately inferred that the human skull has probably not been derived from the usual gorilloid type in which the sutural

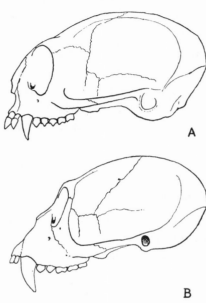

FIG. 15.—LATERAL VIEW OF THE SKULL OF A, MARMOSET, AND B, GIBBON.

Illustrating the general appearance of the simpler types of skull in the Anthropoidea. Note the reduction of the facial region, the forward direction of the orbital apertures, and the relatively large size of the neurocranium in comparison with lower Primates.

pattern is less primitive, but this does not preclude altogether a pithecoid ancestry for the Hominidæ. It should also be noted that it is quite common to find a fronto-temporal articulation among the skulls of primitive human races.

In the New World monkeys the malar is large and articulates with the parietal. This is a specialization which has been avoided in the other groups of the Anthropoidea. The premaxilla is of moderate development and is never reduced to the extent seen in modern lemurs. It usually reaches up along the lateral border of the nasal bone and, in some Old World monkeys, may even extend up as far as the frontal bone.

The lachrymal bone is confined to the orbital cavity (except for a very slight facial extension which may be noted in *Cebus*, *Macacus* and *Papio*), and the lachrymal foramen is within the orbital margin.

The recession of the snout which is foreshadowed to some degree in the lemurs and tarsioids is carried to a further stage in monkeys generally, and there is a corresponding dwindling of the olfactory chambers. The baboons appear to form a conspicuous exception to this general tendency. The projecting muzzle in these animals, however, is to be explained by the fact that in mammals at a corresponding level of cerebral development the brain is *relatively* smaller to the size of the skull in the larger species, while, on the other hand, the jaws and teeth are more powerfully developed in *direct* relation to the body weight. Thus, in the skulls of the larger monkeys and apes the facial skeleton is as a rule more bulky as compared with the size of the neurocranium. In the skulls of small and primitive monkeys the jaws are markedly reduced in proportionate size and herein contrast strongly with the skulls of lower mammals of an equivalent body weight.

The basicranial axis is bent so that the facial part of the skull comes to lie more below the front of the neurocranium than it does in lower Primates. Similarly, the foramen magnum is displaced completely on to the basal aspect of the skull and looks downwards. An exception is seen in the Howler monkey, in which the foramen looks almost directly backwards, but it seems that this displacement is secondary to a change in the poise of the head associated with the prodigious development of the jaws in this creature.

The orbits in the pithecoid skull are large and directed straight forwards, and they are almost completely cut off from the temporal region by a bony wall formed by the malar and alisphenoid (an advanced character which, it has been seen, occurs to a slight degree in *Tarsius*). The os planum of the ethmoid forms a part of the medial wall of the orbit, while the orbital plate of the palatine is restricted in size.

In the New World monkeys the tympanic cavity is distended into an osseous bulla which is formed from the petrous bone, and the entocarotid artery pierces the bulla posteriorly on its ventral surface. There is no persistent stapedial artery as in lower Primates (though this vessel is always present in fœtal stages of ontogeny). The ectotympanic forms a simple ring which is adherent to the outer aspect of the bulla (Fig. 13, C). In the Old World monkeys, apes and Man no definite bulla is developed, and in these forms also, the ectotympanic is produced into a tubular meatus as in the tarsioids (Fig. 9, D).

The mandible in the Anthropoidea shows a progressive tendency towards a shortening of the horizontal ramus and a heightening of the ascending ramus. These changes are clearly related to the recession of the jaws and the bending downwards of the face to take up a position below the front of the neurocranium. In the primitive fossil *Parapithecus* (of Oligocene age) the lower jaw in its general proportions curiously resembles that of the tarsioids, for the horizontal rami diverge considerably from the symphysis. This must have been associated with a relatively wide cranial base and thus indicates an expansion of the brain. In the larger monkeys, however, the rami are almost parallel. This difference is no doubt largely a function of body weight, for, as was pointed out above, in a group of animals of equivalent cerebral status the smaller members have brains which are larger in direct proportion to the size of the body. In the large anthropoid apes the rami also show little divergence, the dental arcade forming an elongated U. In Man, however, the great expansion of the brain has led—independently of the size of the body as a whole—to a great widening of the skull base, and once again the rami of the lower jaw respond by acquiring a V-shaped divergence from the symphysis.

No fossil skulls of the Anthropoidea are known which are more primitive than any of those which may be found among existing forms. Thus we are compelled to rely on the morphological features of the latter in any attempt to construct a hypothetical picture of the cranial form of the earliest and ancestral type of monkey. From this evidence (and assuming for the moment that the Platyrrhine and Catarrhine monkeys are derivatives of a common pithecoid ancestor) it may be supposed that the skull of such a type showed the following features, in respect of which, at least, it certainly could not have been less primitive (though it may very well have been more so):

1. Relatively voluminous and rounded neurocranium with absence of bony crests.
2. Frontal bone forming an extensive part of the cranial roof.
3. Orbits rather large and directed forwards.
4. Lachrymal almost entirely confined to the orbital cavity, and the lachrymal foramen within the orbital margin.
5. The os planum of the ethmoid forming part of the medial wall of the orbit.
6. Marked reduction of the facial region and the nasal cavities.
7. Premaxilla of moderate size.
8. Orbit separated from the temporal region by an almost complete bony wall formed largely by the alisphenoid.
9. Spheno-parietal contact in the temporal fossa.
10. Foramen magnum on the base of the skull and looking almost directly downwards.
11. Prominent and rounded occiput.
12. Marked flexure of the basicranial axis so that the jaws come to be displaced rather below the front part of the brain-case.
13. Tympanic bulla formed from the petrous and perforated by a large entocarotid artery (also a tarsioid character).

14. Absence of the stapedial artery.

15. Ectotympanic element a simple ring applied to the outer surface of the bulla, and no tubular auditory meatus.

It will be observed that a hypothetical skull of this type would be essentially a platyrrhine skull except that it would have avoided a few minor specializations such as the enlarged malar articulating with the parietal bone. Further, it provides a foundation for the elaboration of the more advanced characters of the catarrhine skull. Thus, at least so far as the skull by itself is concerned, there seems no fundamental objection to the derivation of the Catarrhines from a Platyrrhine ancestor.

Among the lower Primates it is clear that the tarsioid skull approximates most closely to that of the Anthropoidea, for it has been shown that some of the highly characteristic pithecoid traits are apparently foreshadowed in the cranial form and structure of the Tarsioidea. Of the tarsioid skulls which are known, however, the cranial characters of *Tarsius*, *Tetonius* and *Necrolemur* are clearly too specialized or have advanced too far in structural elaboration to be entitled to consideration as forerunners of the monkeys.* Thus in these forms there is already a tubular auditory meatus which clearly must have developed independently of the meatus in the higher Anthropoidea, unless, indeed, an independent derivation of the Platyrrhines and Catarrhines from the tarsioids is postulated, the former arising from a primitive tarsioid in which the tubular meatus had not yet appeared, and the latter from a more advanced form with a meatus. This question of a polyphyletic origin of the monkeys will be considered in detail later. So far as the skull is concerned, it seems more

* The fragmentary remains of the skull of *Cænopithecus lemuroides* (from the middle Eocene of Europe), judging from the figures in Stehlin's monograph,[10] indicate that the facial region was remarkably pithecoid in general proportions. This genus is provisionally referred to the Tarsioidea, but the orbit is of moderate dimensions only, and the dentition exhibits certain divergent characters such as the pronounced development of mesostyles in the upper molars.

reasonable to suppose that the tubular meatus arose in the tarsioids as the result of parallelism, the more so since (as emphasized in the previous section) other pithecoid features of the tarsioid skull may be entirely secondary to the highly specialized construction of the orbital region characteristic of this sub-order. It is clear that if the Anthropoidea were derived phylogenetically from a tarsioid ancestor, they must have branched off from the Tarsioidea when the skull as yet manifested none of the peculiar tarsioid modifications of the orbital region, though quite possibly at this stage the orbit was beginning to be shut off from the temporal fossa, and the features of the bulla and associated vessels common to tarsioids and the higher Primates were already indicated. Such a skull type (which might conceivably have formed the morphological basis for the derivation of the specialized tarsioids on the one hand, and the progressive pithecoids on the other) would be of quite a primitive nature, approximating in some degree to that seen in the lemuroid genus *Pronycticebus*—at least as far as the general proportions of the skull and the relative size of the orbits are concerned. This hypothetical type might legitimately be referred to the Tarsioidea even in the absence in their fully acquired form of the typical specializations of this sub-order. Thus, on the basis of skull structure, there seems to be no objection in the way of conceiving a tarsioid ancestry for all the higher Primates.

It has been pointed out in the previous section that the tarsioid skull could hardly have been derived from a skull of the type found in the known Lemuroidea, for the simple reason that the more primitive features in the tympanic region of the Tarsioidea are unlikely to have arisen from the more specialized features of the Lemuriformes and Lorisiformes. The same objections apply to the suggestion that the pithecoid skull evolved from a lemuroid type such as *Adapis* or *Notharctus*. Moreover, to derive the Anthropoidea directly from the Lemuroidea in a scheme of phylogeny in this way would necessarily eliminate the Tarsioidea from a place in the ancestry of higher Primates. On the other hand, it has been recorded above that there are many pithecoid traits in the tarsioid skull which are not to be found among the lemurs, and this surely indicates at least that the Anthropoidea and the Tarsioidea have been derived from an ancestral form with evolutionary potentialities and tendencies different in many ways from the progenitor of the Lemuroidea.

References

1. CLARK, W. E. LE GROS : The Skull of Pronycticebus Gaudryi. Proc. Zool. Soc., 1934.
2. DUCKWORTH, W. H. L. : Morphology and Anthropology. Cambridge, 1915.
3. ELLIOT SMITH, G. : Essays on the Evolution of Man. London, 1927.
4. GRANGER, W., and GREGORY, W. K. : A Revision of the Eocene Primates of the Genus Notharctus. Bull. Amer. Mus. of Nat. Hist., vol. xxxvii., 1917.
5. GREGORY, W. K. : On the Structure and Relations of Notharctus. Mem. Amer. Mus. of Nat. Hist., vol. iii., 1920.
6. GREGORY, W. K. : On the Classification and Phylogeny of the Lemuroidea. Bull. Geol. Soc. Amer., vol. xxvi., 1915.
7. VAN KAMPEN, P. N. : Die Tympanalgegend des Säugethiereschädels. Morph. Jarhb., vol. xxxiv., 1905.
8. KOLLMANN, M. : Études sur les Lémuriens. Mémoires de la Société Linnéenne de Normandie, 1924.
9. MATTHEW, W. D. : Stehlinius, A New Eocene Insectivore. Amer. Mus. Novit., No. 14, 1921.
10. STEHLIN, H. G. : Die Säugethiere des Schweizerischen Eozäns. Abhand. der Schweiz. Paläont. Gesellsch., vols. xxxviii. and xli., 1912 and 1916.
11. WOOD JONES, F. : Man's Place among the Mammals. London, 1929.

CHAPTER IV

THE EVIDENCE OF THE TEETH

No anatomical features have yielded such fruitful evidence in the enquiry into evolutionary development as the characters of the teeth. This is partly due, of course, to the fact that in fossil forms the teeth are usually the best preserved and often the only remains which are available for study. But it is also a fact that, although no doubt there is a broad relation between dental morphology and type of diet, there are many minor variations of the fundamental plan of tooth structure in different mammalian Orders which are hardly to be attributed directly to functional requirements, and this renders them of especial value in the estimation of zoological affinities.

Representatives of palæontology in America have been mainly responsible for the building up of an elaborate system of comparative odontology which forms the foundation of modern research in the dental anatomy of mammals, and in the present chapter free use has been made of these classical studies.

Before embarking upon an enquiry into the early evolution of the teeth in the Primates, it is convenient to consider the nature of the dentition which may be regarded as typical of a primitive and generalized mammal. It is certain that the mammalian dentition has been ultimately derived from a simple reptilian type in which all the teeth are of a simple peg-like form—a homodont type of dentition. Differentiation of function has, however, been associated with a differentiation in the form of the teeth in mammals generally, leading to the heterodont type of dentition which is already to be seen in certain mammal-like reptiles of late Palæozoic times. Thus the tooth series, instead of being morphologically uniform, comes to be divided into the incisor group occupying the anterior end of the jaws (the premaxilla in the upper jaw) and consisting of rather small peg-like teeth often with a cutting edge; the canine or grasping tooth which is single on each side and usually pointed and prominent; the premolar group consisting of uni- or multicuspidate teeth; and most posteriorly the molars or grinding teeth which are multicuspidate, possessing typically three or more cusps, and which, unlike the rest of the dentition, are not preceded by deciduous teeth.

The monumental studies which have in late years been made on the teeth of recent and extinct mammals make it tolerably certain that the generalized type, which gave rise to eutherian mammals as a whole, possessed on each side of the upper and lower jaws three incisors, one canine, four premolars, and three molars. Such a generalized dentition is represented diagrammatically in Fig. 16, and in this form the dentition would be indicated by the following formula :

$$I. \frac{123}{123} \quad C. \frac{1}{1} \quad P. \frac{1234}{1234} \quad M. \frac{123}{123},$$

or more simply and in less detail :

$$3.\ 1.\ 4.\ 3.$$
$$3.\ 1.\ 4.\ 3.$$

In such an ancestral eutherian it is probable that the incisors were small and cylindrical with rounded tips, and showing a slight tendency to slope forwards in conformity with the slope of the facial surface of the premaxilla and the symphysial region of the mandible. The canine is indicated in the figure as a rather short tooth the apex of which is pointed and projects slightly above the level of the neighbouring teeth. That this projection is a primitive feature of the canine is suggested by the fact that it is commonly found in the oldest fossil mammals, including the Pantotheria—a group which may well have given rise to recent mammalian forms according to Simpson[5]—and that a projecting canine is also a typical feature in the heterodont dentition of the mammal-like reptiles from among which it is probable that the whole mammalian phylum has been derived. Some palæontologists, however, maintain that a projecting and pointed canine is a secondary specialisation, and that primitively this tooth is short and blunt (brachydont), forming a morphological transition between the incisor series in front and the premolar series behind. Such a type of canine is found in the fossil anthropoid *Parapithecus* (Fig. 27), and also in the modern human dentition. But, at least in regard to the latter, there is evidence to indicate that it is the brachydont canine which is secondary, for the length of the root in the human tooth and its lateness in eruption (as well as palæontological evidence)

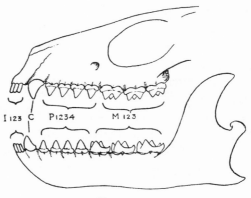

FIG. 16.

Diagram showing the dentition of a hypothetical generalized eutherian mammal, and representing the type from which it is presumed that the dentition of the Primates was originally derived.

seem to indicate that it has been derived phylogenetically from a more powerful and truly "caniniform" canine.

The premolars of a generalized mammal would be of simple form with a single pointed cusp which in the upper teeth is called the *paracone* and in the lower the *protoconid*. The base of the crown is thickened to form a ridge-like basal ring of enamel, the *cingulum*. This part of the tooth, which is well marked in the premolars and molars (and is also to be recognized in the incisors and canine), is of great importance, for it is from the cingulum that certain cusps grow up during the course of evolutionary development, adding thereby to the complexity of the crown of the tooth.

The upper molars of the generalized eutherian type are all tritubercular—that is to say, the crown of each tooth bears three cusps. Of these, one is medial, the *protocone*, and two are lateral, the *paracone* in front and the *metacone* behind. Each tooth, moreover, is anchored to the jaw by three roots, two lateral and one medial. It will be observed that the three cusps form the points of a triangle on the crown of the tooth ; this formation is called the *trigone*. It is from such a tri-tubercular or three-cusped tooth that the variously compli-cated molars of all recent mammalian Orders have probably been derived phylogenetically (Fig. 17). This, at least, is the conclusion which follows from the evidence adduced by Cope and Osborn in the elaboration of their well-known theory of trituberculy, and although this theory has been modified in minor details since its first inception, in the main it may be regarded as an established principle which has been confirmed extensively by recent palæontological studies. An alternative theory is that of multituberculy. Adherents of this theory (which are to-day few in number) would maintain that prim-itively the mammalian molar possessed a large number of small cusps—as seen, for instance, in the Order Multitubercu-lata of late Mesozoic and early Tertiary times—and that, by the elimination of varying numbers of these cusps, the more simple molar patterns of recent mammals have been derived. There are many objections to this theory, however, and even were it true in part for mammals generally, it can hardly be disputed that it was a tritubercular type of molar which formed the basis for the evolutionary development of the molar patterns found in Primates. As will be shown later, in the more primitive Primates the molars approach closely to the tritubercular form postulated by the Cope-Osborn theory.

The crowns of the lower molars of the hypothetical general-ized eutherian mammal depicted in Fig. 16 are composed of two parts, each supported by a root. Anteriorly is the *trigonid*, showing three cusps of which one, the *protoconid*, is laterally placed, while the other two, the *paraconid* and *metaconid*, are medial. The trigonid appears to be analogous to the trigone of the upper molars, and the cusps are named in consonance with this conception. The posterior part of the crown primitively forms a narrow low heel or *talonid*, the surface of which is lower than the surface of the trigonid and is hollowed out to form a depression, the talonid basin. Into this depression, the tip of the protocone of the upper molar fits in occlusion of the teeth. Two small cusps appear on the medial and lateral margins of the talonid basin and are termed respectively the *entoconid* and the *hypoconid*. In the progressive differentiation of the lower molar an additional cusp, the *hypoconulid*, may be formed at the posterior end of the talonid.

Having indicated the nature of the dentition in a generalized eutherian mammal such as might represent the group from which the Primates (and other eutherian Orders) have been ultimately derived, it is convenient next to give a general idea of the prevailing evolutionary tendencies which the teeth show in the Primate series.

The incisors are in almost every known Primate reduced in number, for there are usually but two in each side of the upper and lower jaws. Notable exceptions occur in the fossil tarsioid genus *Omomys*, in which the original number of incisors is retained in the lower jaw, and in *Pseudoloris*, in which there are three upper incisors. In the Lemuroidea the upper incisors become much reduced in size, and in some (*e.g.*, *Megaladapis*) they vanish altogether, while the lower incisors in all recent genera are markedly specialized (Fig. 20, B and C), becoming so procumbent as to lie almost horizontal (except in the aberrant *Chiromys*). In the Anthropoidea the incisors commonly assume a spatulate form with a relatively straight cutting edge, especially the upper teeth.

The canines in many cases hypertrophy to form sharp

and powerful dagger-like teeth, as, for instance, in *Notharctus venticolus*, in the upper canines of recent lemurs, in the larger monkeys such as the baboons, and in the anthropoid apes. On the other hand, they may get taken over into the incisor series, becoming incisiform in shape, as in the specialized lower dentition of recent Lemuroidea and in the modern types of Man.

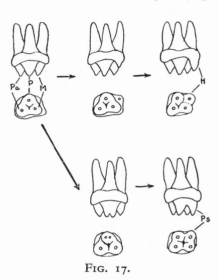

FIG. 17.

Diagram showing the manner in which the primitive tritubercular upper molar may become converted into a quadritubercular tooth by the formation of a true hypocone which grows up from the cingulum at the base of the tooth (above), or a pseudohypocone which arises by fission of the protocone (below). In the side views the teeth are represented from the medial aspect.

Pa, Paracone; P, Protocone; M, Metacone; H, Hypocone; Ps, Pseudohypocone.

The premolars in the Primates show a tendency (1) to diminution in size of the anterior two and a relative enlargement of the third and fourth, and (2) to a disappearance of the first premolar and later of the second premolar also. Thus in the Indrisidæ among the lemurs, and in all Catarrhine monkeys, apes, and Man, there are only two premolars. Moreover, the third and fourth premolars—as in other mammalian Orders —tend to become progressively complicated by the upgrowth of cusps from the cingulum, assuming the appearance and characteristics of molars. This process is termed " molarization," and it forms an essential part of the premolar-analogy theory of Wortmann, who argues very convincingly that these changes indicate the manner in which the primitive tritubercular molar teeth of mammals were originally derived from a simple cone-shaped tooth with a single cusp.[13] In the process of molarization of the upper premolars the original single cusp, the paracone, becomes divided into two, the paracone and metacone, while a third cusp, the protocone, springs up as a spur from the internal cingulum. In many lower Primates these changes have proceeded further than in the higher Primates, and in the Anthropoidea the premolars are commonly bicuspid (the medial cusp usually being termed the deutero-cone). It is, indeed, a general characteristic of the Tarsioidea and the Anthropoidea—in contrast to what appears to be a definite lemuroid tendency—that the pre-molars have almost entirely avoided this specialized feature of molari-zation.

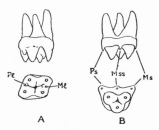

A   B

FIG. 18.

Diagram representing the further elaboration which may occur in the crown of the upper molar by the appearance of A, proto-conule and metaconule, or B, "styles," which spring up on the external cingulum.

Pl, Protoconule; Ml, Meta-conule; Ps, Parastyle; Ms, Metastyle; Mss, Meso-style.

The upper molars in all three sub-orders of the Primates show a progressive tendency to assume a quadritubercular form by the development of a fourth cusp situated postero-internally on the crown. It is of the highest importance to realize that this fourth cusp may be developed in two distinct ways. Usually it springs up from the internal cingulum as shown diagramatically in Fig. 17, and is termed the *hypocone*. In the dentition of primitive Primates the hypocone is either not developed, or is small and (to use a term suggested by Abel[13]) orimentary. In more advanced types it may become equal in proportions to the cusps of the original trigone. A fourth cusp may arise, however, by a splitting or dichotomy of the protocone, and in this case it is termed a *pseudohypocone*. This type of quadritubercular molar is found in the Noth-

arctinæ and Plesiadapidæ, and, in the mode of origin of the
pseudo-hypocone, it resembles somewhat the fundamental
molar pattern of certain ungulates (*e.g.* Perissodactyla).[3] The
upper molars may become further complicated by the addition
of more cusps, for instance a *protoconule* situated between the
protocone and paracone, and a *metaconule* between the metacone and hypocone, or small cuspules may spring up from the
external cingulum in front (*parastyle*), posteriorly (*metastyle*),
or in the middle of the outer margin of the tooth (*mesostyle*)

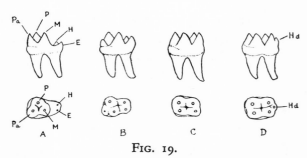

FIG. 19.

Diagram showing the manner in which the primitive tuberculo-sectorial
lower molar, A, may become converted into a quadritubercular
tooth. In B, the paraconid is shrinking, and the entoconid and
hypoconid are becoming elevated. In C, the paraconid has disappeared, and the four remaining cusps are equal in size. In
D, an additional cusp, the hypoconulid, is shown posteriorly.
The side views of the teeth are represented from the medial aspect.

Pa, Paraconid; P, Protoconid; M, Metaconid; H, Hypoconid;
E, Entoconid; Hd, Hypoconulid.

(Fig. 18). This development of "styles" resembles a
common feature seen in the dentition of lipotyphlous insectivores. Lastly, the cusps of the upper molars may become connected by ridges such as the *protoloph* (linking the
protocone with the paracone) and the *metaloph* (linking the
metacone with the hypocone).

In the lower molars a quadritubercular and quadrilateral
crown is developed as an evolutionary tendency in the
Primates in rather a different way. The paraconid becomes
progressively diminished until it finally disappears altogether,

while the protoconid comes to lie more directly lateral to the metaconid. At the same time the talonid widens, its two main cusps, the entoconid and hypoconid, become larger, and the talonid basin fills up. Eventually, the whole talonid is raised up to the level of the trigonid (Fig. 19). A fifth cusp, the hypoconulid, is frequently formed at the posterior margin of the crown of the tooth rather towards the outer side, and a sixth cusp, the *tuberculum sextum*, may appear medial to this (*e.g.*, in *Dryopithecus* and *Gorilla*).

With this general conception of the evolutionary trend of the dentition in Primates as a whole, attention may now be turned to a consideration in more detail of the changes which occur in the teeth of the main subdivisions of this group, and of the inferences which may be drawn therefrom regarding their phylogenetic differentiation.

Lemuroidea

In general, the dentition is more specialized in the Lemuroidea than in the tarsioids or Anthropoidea. It has already been pointed out that in all living lemurs the incisors are peculiarly modified. Except in the Aye-aye (in which there is an extreme specialization of a different type), the upper incisors are shrunken in size, while the lower incisors are very procumbent and styliform. The latter project forwards almost horizontally and with the modified incisiform canine form a curiously comb-like apparatus which the animals actually use for combing the fur (Fig. 20, C). Gregory has suggested that this displacement of the lower incisors is secondary to an enlargement of the tongue* and the development of the sublingua. It is more likely, however, that at

---

* Gregory's implication that in the lemurs the tongue is disproportionately large is perhaps based upon the fact that he examined a preserved specimen, for the tongue tends to swell rather rapidly after death and under the action of certain fixatives.

least the latter feature is conditioned by the conformation of the incisors (*vide infra*, p. 193). The shrinkage of the upper incisors is associated with a reduction of the premaxilla, and the incisors of either side become separated by a comparatively

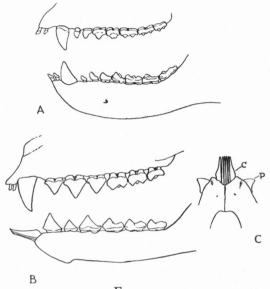

FIG. 20.

Lateral view of dentition of A, *Notharctus crassus* × $\frac{2}{3}$ (after W. K. Gregory), dental formula $\frac{2 \cdot 1 \cdot 4 \cdot 3}{2 \cdot 1 \cdot 4 \cdot 3}$; B, *Lemur varius* × 1, dental formula $\frac{2 \cdot 1 \cdot 3 \cdot 3}{2 \cdot 1 \cdot 3 \cdot 3}$. The lower front teeth of *Lemur varius* are also shown in C, as viewed from below.

Note how, in the modern lemur, the lower incisors become markedly procumbent, the canine is taken over into the incisor series, and the most anterior lower premolar assumes the form of a canine.

C, Canine; P, Premolar.

wide interval. In *Lepilemur* and in the extinct *Megaladapis* the upper incisors (so far as the permanent dentition is concerned) have disappeared altogether.*

* On the other hand, in the Pleistocene genus *Archæolemur* these teeth attained to a relatively large size, and acquired a spatulate form as in the Anthropoidea.

In no member of the Lemuroidea of which the dentition is sufficiently known is the primitive number of the incisors (three on each side) retained. In the fossil Adapidæ (*Adapis, Notharctus*) the incisors are two in number on each side of the jaw, and in their shape and position they preserve to a considerable degree the generalized features of the mammalian incisors, showing none of the typical lemuroid specializations which all recent lemurs exhibit. Gregory[2] states that the upper incisors of *Adapis* are rather more specialized than those of *Notharctus* in that they have a chisel-like cutting margin, while the upper incisors of *Notharctus* " were already beginning to assume somewhat the compressed shape which is retained in certain modern lemurs (*Cheirogaleus*)." The main point of interest here lies in the fact that the Adapidæ (as shown by their skull structure) are related essentially to the Lemuriformes of which group they are early representatives, and it seems evident, therefore, that the Lemuriformes and Lorisiformes were derived from an ancestral form of lemuroid before the characteristic specialization of the front teeth (incisors and lower canines) had developed. Thus it appears that this extreme modification has arisen independently in these two main divergent groups of the Lemuroidea. This conclusion inevitably leads to the conception of an orthogenetic trend of evolution dependent upon an inherent tendency in the common progenitor to the production of similar features in divergent groups of descendants. For, although clearly the front teeth of modern lemurs are used similarly in different groups for toilet purposes, there is no demonstrable reason for supposing that the characters of the furry coat in these animals are in any way so peculiar that they *inevitably* demand the elaboration of such a dental comb.

The upper canines in many living lemurs are large and tusk-like, a modification which was carried to an extreme in *Megaladapis*. In the primitive Adapidæ they show some variation which may (according to Stehlin) depend upon sexual differences. In *Notharctus osborni*, and in presumed female specimens of *Adapis parisiensis*, the tooth is relatively small and but slightly projecting. In all modern lemurs the lower canines have become relegated to the incisor series and are markedly procumbent (Fig. 20.) This specialization was not present in the Adapidæ, as already mentioned.

The premolars preserve their original number in the Adapidæ and certain plesiadapids, but in all other families of the Lemuroidea they have been reduced, usually (by the disappearance of the first of the series) to three. In the Indrisidæ the second premolar has also vanished, so that in this family there are only two premolars left on each side (as in the Catarrhines). In most of the recent forms (*e.g., Lemur, Loris, Perodicticus*) the upper premolars have no more than two cusps, the main cusp or paracone, and a low internal basal cusp, the deuterocone. In some genera, however, the last premolar has undergone some degree of molarization (*e.g., Galago*), with the appearance of a metacone and an orimentary hypocone. This condition is to be regarded as a more advanced stage in the evolutionary differentiation of the premolar series, and thus it may be reasonably inferred that the ancestral type of lemuroid possessed upper premolars at least as primitive as those of (for example) *Lemur*. On the other hand, the argument may be raised that the simpler premolars in the Lemuridæ are the result of a secondary degeneration from molariform precursors, but for this conception—at any rate as far as concerns the larger species—there is no strong evidence. Since in the fossil *Adapis* and *Notharctus* the last premolar is already molarized to some degree, showing two external cusps, we may agree with Stehlin that the dentition of these genera would hardly have given rise to that of the modern lemuroids (see Fig. 21). However, in *Pronycticebus* (which is a primitive adapid) the premolars are relatively unelaborated, and this suggests the possibility of the derivation of recent lemurs phylogenetically from at least the earlier representatives of the Adapidæ. In some of the earlier Notharctinæ also (*e.g., Pelycodus ralstoni* and *Pelycodus trigonodus*),

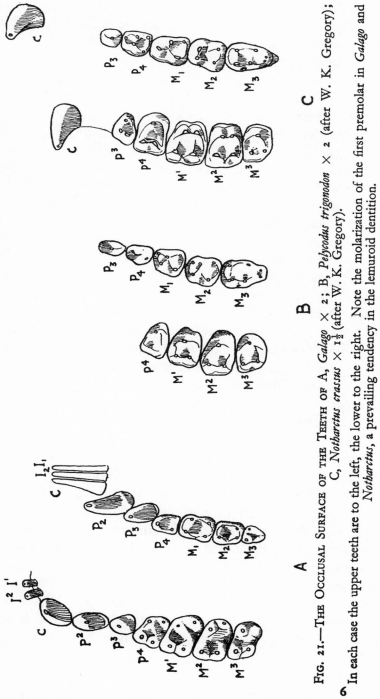

FIG. 21.—THE OCCLUSAL SURFACE OF THE TEETH OF A, *Galago* × 2; B, *Pelycodus trigonodon* × 2 (after W. K. Gregory); C, *Notharctus crassus* × 1½ (after W. K. Gregory).

In each case the upper teeth are to the left, the lower to the right. Note the molarization of the first premolar in *Galago* and *Notharctus*, a prevailing tendency in the lemuroid dentition.

the premolars have not assumed a tritubercular form, but in this case (as will be seen later) there are other characters which definitely debar all the Notharctinæ from a place in the ancestry of modern lemuroids.

The most anterior lower premolar in all modern lemurs is specialized to a varying degree by assuming the function of the true canine which—as recorded above—has been relegated to the incisor series.   It thus becomes caniniform or subcaniniform, relatively large and pointed, and works in opposition to the upper canine. A similar change, but less in degree, may affect the anterior premolar of the upper series, which is functionally related to the caniniform lower premolar (*e.g.*, in *Perodicticus*).   The posterior lower premolars of the Lemuroidea as a whole are fairly simple, though the last of the series usually shows a small talonid.   As in the upper series, the last lower premolar in *Adapis* and *Notharctus* is somewhat more molariform than it is in many recent genera.   Lastly, it may be noted that in certain fossil Malagasy lemurs of Pleistocene age (*Archæolemur, Bradylemur*) the premolars all show a remarkable specialization in the formation of a continuous sharp cutting edge by the lateral compression of their crowns.

The upper molars in the Lemuridæ are of typical tritubercular pattern.   In *Lemur* there is a slight indication of a true (cingulum) hypocone.   Among the Lorisiformes, *Perodicticus* shows a minute hypocone, while in *Nycticebus* and *Loris* this cusp may be quite conspicuous, and in *Galago* the quadritubercular form of the molar is well developed (Fig. 21, A).   In the Eocene representatives of the Lemuroidea a true hypocone is already present in the molars of the Adapinæ (*Adapis* and *Pronycticebus*).   In *Notharctus* the molars are also quadritubercular, but, as in all the Notharctinæ, the fourth cusp is a pseudohypocone produced by fission of the protocone (Fig. 21, C). This unusual feature must be regarded as an aberrant specialization which cannot be supposed to have provided a basis for the evolutionary origin of modern lemurs or, indeed, of higher Primates.   Gregory has sought to minimize the significance of this mode of origin of the fourth cusp. He argues that it is merely a secondary character following a different type of movement at the temporo-mandibular joint. But if this articulation is so specialized in the Notharctinæ as to determine such an unusual origin of the hypocone, then this character in itself is evidence of the aberrant nature of the sub-family and would equally preclude the whole group from participation in the ancestry of later forms in which the fourth cusp is a true cingulum hypocone. It is true that in the earlier forms of *Pelycodus* (primitive precursors of *Notharctus*) the upper molars are tritubercular (the pseudohypocone is not present in *P. ralstoni* or *P. trigonodus* and is just beginning to appear in *P. frugivorus*), but the whole sequence of fossil remains of this genus illustrates in a most striking way a progressive tendency towards the full development of the pseudohypocone as seen in *Notharctus* itself. Moreover, in *Notharctus* and in the later stages of the evolutionary development of *Pelycodus*, the upper molars are further complicated by the appearance of mesostyles, another specialization which serves to indicate the divergent nature of the Notharctinæ. It may be agreed with Stehlin, therefore, that the Notharctinæ form a group by themselves, sharply distinguished from the Adapinæ and from all recent lemuroids in the characters of their molars.

The lower molars in the living species of the Lemuroidea lack the paraconid, the protoconid lies opposite the metaconid, and the talonid basin tends to be rather broadened out (except in the last molar) and raised in level. The more generalized condition of the tuberculo-sectorial tooth is seen in *Notharctus*, in which a small paraconid is present, at least in the first molar. In the basal Eocene representatives of *Pelycodus* (e.g., *P. ralstoni*) the paraconid is well developed in all three molars, and the crowns of the latter approximate closely to the primitive placental type. The gradual disappearance of the paraconid in the Notharctinæ is remarkably well demonstrated by the palæontological sequence available. In the known forms of *Adapis* the paraconid had already disappeared.

Thus far we have not referred to the anomalous dentition of the Aye-aye (*Chiromys*). In this very aberrant lemur the front pairs of teeth in the upper and lower jaws (which are probably incisor teeth) are extraordinarily modified so as closely to resemble the incisor teeth of rodents (Fig. 12, B). They have a persistent pulp, show continuous growth, and are covered only on their anterior surface by enamel. The back teeth, on the other hand, have undergone a considerable degree of degeneration, and there remain, in a vestigial condition, one premolar and three molars in the upper jaw and three molars in the mandible. The chief interest of *Chiromys* in the present discussion lies in its possible relationship to the extinct Plesiadapidæ. Fragmentary remains of this group have been collected from Palæocene and Eocene deposits of Europe and America. As longa go as 1880 Lémoine drew attention to the fact that the progressive modifications of the incisors in the plesiadapids provide a series which leads by gradual steps to the specialized incisor teeth of the modern Aye-aye,* and in this view he was followed by Schlosser and Osborn. Matthew preferred to allocate the group to the Menotyphla (tree-shrews). However, such authorities as Stehlin, Teilhard de Chardin and Abel have recently emphasized again the remarkable parallelism between the Plesiadapidæ and *Chiromys*, and other modern authors—even if they hesitate somewhat at such an association—regard them as definitely Primates. But if the latter assumption is granted, then the evolutionary tendency shown in the incisors of fossil plesiadapids (bearing also in mind the geological phase at which this tendency was becoming manifest) does indeed make it very highly probable that the group as a whole may really be closely related to the direct ancestral stock of *Chiromys*.

In the upper Eocene genus *Stehlinella* the incisors had already attained to a large size, and the enamel layer in the lower incisor was confined to the anterior surface of the tooth

---

* Even though the *known* Plesiadapidæ were too specialized to have formed a basis for the direct ancestry of the Aye-aye.

(Fig. 12, A). Besides the three molars in the maxilla and mandible, this form had two upper and one (very specialized) lower premolar. The canines were absent as in the Aye-aye. In *Trogolemur* (middle Eocene), which is known only by a part of the lower jaw, there were three lower premolars. In *Heterohyus* (=*Necrosorex*) the dental formula was $\frac{2 \cdot 1 \cdot 1 \cdot 3}{1 \cdot 0 \cdot 2 \cdot 3}$; the first lower incisor was large and covered only on its anterior surface by enamel, but apparently it did not have a persistent pulp. In *Phenacolemur* (upper Palæocene and lower Eocene) there were two upper and two lower premolars, the second lower premolar being somewhat specialized as in the tarsioid genus *Carpolestes*, while the lower incisors were covered both front and back by enamel. In the Palæocene genus *Pronothodectes* the premolars were four in number, and thus correspond to the primitive placental condition. Lastly, in *Plesiadapis*, in which the anterior incisors, though large, were completely covered on the crown by enamel and had not attained to the specialized condition seen in later forms, the dental formula is given as $\frac{2 \cdot 1 \cdot 3 \cdot 3}{2 \cdot 0 \cdot 2 - 3 \cdot 3}$ (Abel). Even in this primitive Primate, it should be noted, specialization of the teeth had proceeded relatively far, for not only are the incisors affected and the lower canine lost, but the last two premolars (P^3 and P^4) are molariform with three cusps, while the molars have become quadritubercular and show a distinct mesostyle. It is important to realize that the fourth cusp of the plesiadapid molar is a pseudohypocone (Stehlin), a feature in which the Plesiadapidæ are associated with the Notharctinæ (and incidentally show no affinity with the tree-shrews).

It is to be presumed that if the Plesiadapidæ are closely related to *Chiromys*, they must be regarded as early representatives of the Lemuriformes (of which the Chiromyidæ form a subdivision), and thus they must have been derived phylogenetically from a group which was also ancestral to the early

Adapidæ. The fossil remains of the Plesiadapidæ indicate that such a precursor was certainly a very small mammal with an extremely primitive dentition.

If we now survey the dental characters of the Lemuroidea in general, we may draw the following inferences. In all modern lemurs (except *Chiromys*) the teeth are greatly specialized in the dwindling of the upper incisors and the porrect and styliform condition of the lower incisors, the relegation of the lower canine to the incisor series, and the adoption of a caniniform shape and function by the anterior lower premolar, P_2 (or, in the case of the Indrisidæ, P_3). In the Pleistocene lemurs of Madagascar more extreme modifications are seen in the canines and premolars, in the appearance of selenodont and bilophodont molars and (in the Archæolemuridæ) in the shrinkage of the whole molar series. The fossil Adapidæ, however, demonstrate the earlier phases in the evolution of the lemuroid dentition. From these it is seen that quite early the lemuroid incisors were reduced to two, while these and the canine were of a generalized eutherian type, and the premolars retained their primitive number at least into the upper Eocene. The Notharctinæ exhibit an unusual molar specialization in the development of a pseudohypocone, and thus, it appears, can hardly occupy a position in the direct line of descent of modern lemurs. The upper premolars of known specimens of *Adapis* had already undergone a degree of molarization which has not been attained by all recent lemurs, and thus, again, it is difficult to regard them in a direct ancestral relation to the latter. It is possible, however, that in more primitive representatives of this genus the premolars were of simpler construction, as Gregory has suggested in the case of *Adapis sciureus* (in which P^4 is missing from the fossil jaw). Certainly the premolars of the primitive adapid *Pronycticebus* were sufficiently generalized to have provided a basis for the derivation of the modern lemuroid dentition. In this genus, however, the hypocones in the molars are distinct, and, indeed, the palæontological evidence indicates that in the

Adapidæ generally this cusp was developed very early in the upper molars and rapidly became very large (as Dr. G. G. Simpson has pointed out to me). It may be surmised that among the earliest Adapidæ there probably occurred forms in which the dentition was sufficiently generalized to have provided a morphological basis for the evolution of the whole lemuroid group, but, since the skull structure demonstrates that the Adapidæ were already committed to the lemuriform (in contrast to the lorisiform) trend of development, it is clear that these two divergent groups separated before the adapid stage had been reached during phylogeny. The evidence of the Plesiadapidæ suggests still more emphatically how early this separation probably occurred. On the evidence so far adduced, in fact, it seems that the ancestral type from which the dentition of the Lemuroidea was derived probably possessed tooth characters almost identical with those postulated for the generalized eutherian mammal (*vide supra*), except that the incisors may already have been reduced from three to two. Incidentally, the fact that the dentition in the Lemuroidea was able during the course of evolution to become modified in such aberrant directions as indicated in the modern lemurs generally, in the Aye-aye, and in the Archæolemuridæ, is a further corroboration of the conception that the common ancestor of these forms must certainly have been a remarkably generalized mammal endowed with the widest potentialities.

Tarsioidea

As a whole, the dentition of *Tarsius* is more primitive than that of any of the recent Lemuroidea, and indeed more so than that of most extinct tarsioids. In this living form the dental formula is $\frac{2 \cdot 1 \cdot 3 \cdot 3}{1 \cdot 1 \cdot 3 \cdot 3}$ (Fig. 22, A). The incisors are implanted vertically in the upper jaw, and in the mandible show a slight procumbency only. They show specializations in the reduction from the primitive number and in the enlargement of the first upper incisor. In the lower Eocene tarsioid *Omomys belgicus* there were three lower incisors, as in the generalized eutherian dentition. In *Hemiacodon, Washakius,* and *Anaptomorphus* one of these had been lost, while in *Necrolemur* and

Tetonius the lower incisors had all disappeared, their place being taken functionally by enlarged and highly specialized lower canines. It may be noted, also, that in *Omomys* (Fig. 22, B) and *Hemiacodon* the lower incisors were markedly procumbent.

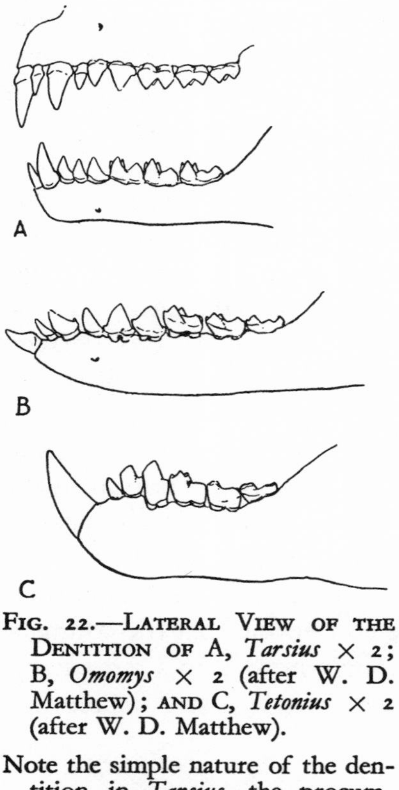

In *Tarsius* the canines are sharp and somewhat projecting, but in general not notably specialized. The upper premolars are of a simple form, the most anterior being very small and unicuspidate, while the two posterior each have a prominent main cusp and a small internal basal cusp (deuterocone). There is no molarization of the last premolar, so that the premolar series stands out in abrupt contrast with the molar series. The lower premolars are correspondingly simple, the main cusp or protoconid being supplemented only by a very small and insignificant internal basal cusp. In some extinct tarsioids the premolars show various degrees of specialization. Thus, in *Tetonius* (Fig. 14, B), *Uintanius, Absarokius* and others the last upper premolars are conspicuously

FIG. 22.—LATERAL VIEW OF THE DENTITION OF A, *Tarsius* × 2; B, *Omomys* × 2 (after W. D. Matthew); AND C, *Tetonius* × 2 (after W. D. Matthew).

Note the simple nature of the dentition in *Tarsius,* the procumbency of the incisors in *Omomys,* and the absence of lower incisors and the highly specialized canine in *Tetonius.*

enlarged. In *Carpodaptes* (of which only the lower teeth are known) the last premolar is highly modified and attains to an unusual size. In *Carpolestes* the specialization of this tooth is still more remarkable, for in this genus it comes to bear a resemblance—even though superficial—to the serrated shearing teeth of the Plagiaulacoidea, a sub-order of the Mesozoic multituberculates. It is interesting to note that some fossil genera (*e.g.*, *Necrolemur*, *Pseudoloris*, *Omomys*, *Carpodaptes*) retain

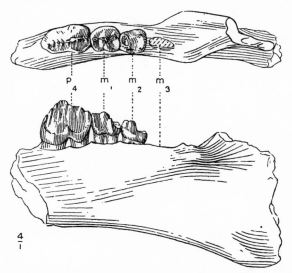

FIG. 23.—THE MANDIBULAR FRAGMENT AND TEETH OF *Carpolestes nigridens*, A *Palæocene* TARSIOID, SHOWING THE EXTREME SPECIALIZATION OF THE LAST PREMOLAR TOOTH × 4. (G. G. Simpson, *Amer. Mus.* Nov., 1928).

the four lower premolar teeth of the generalized eutherian dentition, though it must be emphasized that the tarsioids as a whole (and in contrast with the adapids) show a tendency to a reduction of the premolar series even so early as the middle Palæocene.

The upper molars of *Tarsius* (Fig. 24, A) are of the simple tritubercular pattern, with the addition of a minute protoconule and a thickening of the cingulum in the position where the hypocone develops in quadritubercular molars. The

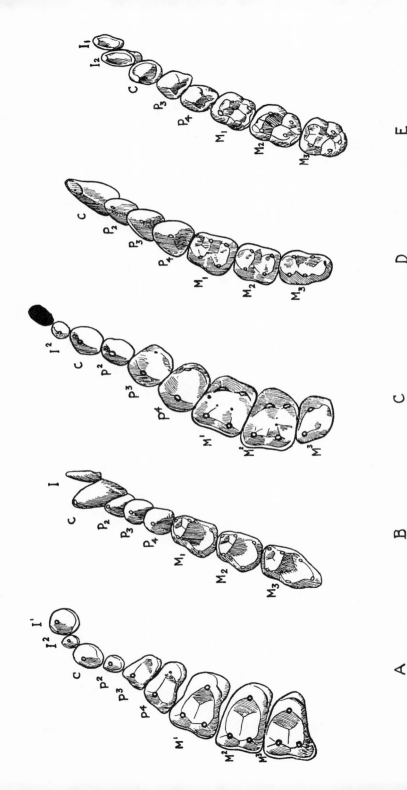

FIG. 24.—THE OCCLUSAL SURFACE OF A, THE UPPER TEETH OF *Tarsius* × 5; B, THE LOWER TEETH OF *Tarsius* × 5; C, THE UPPER TEETH OF *Necrolemur* × 4; D, THE LOWER TEETH OF *Necrolemur* × 4; AND E, THE LOWER TEETH OF *Parapithecus* × 3. C and D, after Stehlin; E, after Schlosser.

lower molars are likewise of a generalized type, with a complete trigonid bearing the three cusps, protoconid, paraconid, and metaconid, and a low hollowed-out talonid in the margins of which are the entoconid, hypoconid, and a small hypoconulid (Fig. 24, B). In some of the extinct tarsioids the molars are more highly developed, showing a significant approach in their structure to the molar teeth of the higher Primates. In *Necrolemur*, for instance, there is a well-developed fourth cusp, a true hypocone, in the upper molars, but these teeth are further complicated by the development of a conspicuous protoconule and metaconule as well as a small parastyle (Fig. 24, C). In the lower molars of this fossil genus the talonid is raised up almost to the level of the trigonid, while the paraconid has disappeared in the two posterior teeth (Fig. 24, D). In *Microchœrus* the crowns of the premolars and the molars show a specialization in the development of a number of secondary cuspules, so that Abel has compared the teeth of this extinct tarsioid in their elaboration with those of the orang. Moreover, this genus shows a mesostyle in the upper molars which is not present in *Necrolemur*, and the paraconid has virtually disappeared in the lower molars. In *Anchomomys* there is a hypocone in the first two upper molars (but not in the last), and in the lower molars the paraconid is rudimentary. In *Cænopithecus* the upper molars are also quadritubercular, with the addition of a mesostyle and a well-developed protoconule, the paraconid being absent in the lower molars.

It will be apparent from the foregoing account that, in comparison with the dentition of fossil tarsioids, that of the modern tarsier is remarkably primitive. It is equally clear that the sub-order as a whole has been derived from a common progenitor which still preserved the primitive eutherian dental formula. Moreover, in this ancestral form, the structure and conformation of the individual teeth must have approximated closely to the generalized mammalian condition. Thus the incisors presumably showed no marked procumbency, the canines were slightly projecting and erect, the premolars exhibited at the most only a small deuterocone in addition to the paracone, while the molars were of the simple tritubercular and tuberculo-sectorial type. During the evolutionary radiation of the tarsioids in Palæocene and Eocene times, there were manifested tendencies towards various specializations such as a progressive reduction of the incisors, an enlargement and procumbency of the canines, the elaboration and hypertrophy of the last premolar (highly characteristic of the American genera), and the complication of the crowns of the molars. The latter feature is of special interest, for in some genera the molar pattern shows an approximation to that of the Anthropoidea. In the tarsioids generally, also, the contrast in structure between the premolar and molar series associated with the absence of molarization of the former, and the tendency (in the European fossil forms) for the last two premolars to resemble each other closely in size and shape, offer a further point of similarity with the dentition of higher Primates.

In the problem of the evolutionary origin of the tarsioid dentition, it will be observed that the Lemuroidea (so far as they are at present known) hardly come in for consideration. All recent lemurs are too specialized in their dentition, especially in the incisor and canine series, as representatives of groups which may have given rise to the Tarsioidea. Among the extinct forms, *Adapis* must be excluded because, *inter alia*, of the specialization of the last premolars, *Pronycticebus* because of the development of a distinct hypocone, and the Notharctinæ because of the anomalous manner in which the quadritubercular molar is developed. If the Tarsioidea can rightly be said to have arisen phylogenetically from the Lemuroidea, they must have separated from the latter at a stage when the dentition was still of such a generalized type that it would show nothing distinctive by which it could be definitely labelled " lemuroid."

Anthropoidea

The Platyrrhine and Catarrhine monkeys differ significantly in their dental formula. In the former this is $\dfrac{2 \cdot 1 \cdot 3 \cdot 3}{2 \cdot 1 \cdot 3 \cdot 3}$. The Hapalidæ provide an exception, however, in that the upper and lower last molars have been lost. In the Catarrhine monkeys the formula is $\dfrac{2 \cdot 1 \cdot 2 \cdot 3}{2 \cdot 1 \cdot 2 \cdot 3}$, which is identical with that of the anthropoid apes and Man.

In the preservation of three premolars, the New World monkeys are evidently the more generalized, and Gregory (whose studies in comparative odontology are almost unrivalled) believes that the genus *Callicebus* exhibits the most primitive dentition of any of the Platyrrhines. In this animal the incisor teeth have assumed a moderately spatulate form, especially the central upper incisors. The canines are slightly projecting, and there is no diastema in relation to them. The three upper premolars have broad

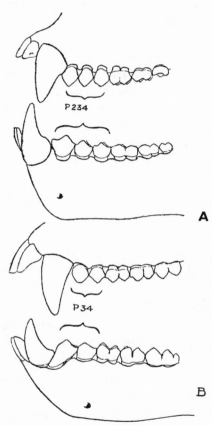

P234

A

P34

B

FIG. 25.—LATERAL VIEW OF THE DENTITION OF A, A NEW WORLD MONKEY, *Cebus* × $\frac{4}{3}$; AND B, AN OLD WORLD MONKEY, *Macacus* × $\frac{3}{4}$.

crowns with two cusps, the paracone and deuterocone, the former somewhat more prominent than the latter. The most

anterior premolar is rather smaller than the posterior teeth.
The lower premolars are very simple, the deuterocone in the
anterior two teeth being indicated by little more than a thickening of the internal cingulum.   The upper molars are quadritubercular, the cusps are sub-equal in size, and the hypocone
is evidently formed as an upgrowth from the cingulum,
being separated from the protocone by a conspicuous fissure.
The anterior and posterior pairs of cusps lie each in an approximately transverse plane.   The last molar is retrogressive and
much smaller than the second.   This atrophy of the last
molar seems to be an expression of a definite evolutionary
tendency in the Platyrrhines as a group, reaching its culmination in the marmosets in which the tooth finally disappears
altogether.   The lower molars in *Callicebus* are also quadritubercular, the talonid being raised up to the level of the
remains of the trigonid.   There is no paraconid.

In the Old World monkeys, the Cercopithecidæ, the two
premolars are typically bicuspidate, though in some forms the
anterior lower premolars are enlarged and specialized in
association with the development of a prominent upper
canine.   The cercopithecid molars also show a specialization
in the linking up of the anterior and posterior pairs of cusps
by transverse ridges, producing a bilophodont form of tooth
(Fig. 26, B).   This bilophodonty (it should be noted) would
debar the Cercopithecidæ from consideration as a group from
which the Simiidæ and the Hominidæ might have been derived,
for in these two families the molars preserve a more primitive
structure.   It may also be remarked that in the Cercopithecidæ the hypoconulid tends to disappear, a divergent
modification which is avoided in the anthropomorphous
apes.

Of the fossil genera of monkeys, two examples provide
valuable evidence in regard to the evolutionary origin of the
Anthropoidea.   Of these, *Apidium*—known by a part of the
lower dentition found in lower Oligocene deposits of Egypt
—represents a form which (according to Gregory) marks the

commencing separation of the specialized cercopithecids from the progenitors of the anthropoid apes and Man, a separation which had certainly become complete well before the Pliocene, as indicated by the fossil monkey *Oreopithecus*. *Apidium* provides interesting features in the sharp structural differentiation of the premolar from the molar series, in the fact that the last lower premolar is only " incipiently bicuspid," and in the retention of the paraconid in the first molar and the presence of a well-developed hypoconulid. The other fossil form is

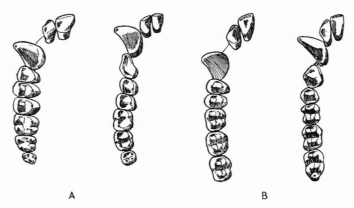

A   B

Fig. 26.—The Occlusal Surface of the Teeth of A, *Cebus* × $\frac{4}{3}$; and B, *Macacus* × $\frac{3}{4}$.

In each case the upper teeth are to the left, the lower to the right. Note the marked degree of bilophodonty in the molar teeth of *Macacus*, a distinctive cercopithecid character.

Parapithecus, which also comes from the lower Oligocene of Egypt and which is known only by the lower jaw and teeth. The dental formula of *Parapithecus* is identical with that of the Old World Anthropoidea. The teeth are extraordinarily generalized, and there is good reason for believing that they represent a type of dentition from which that of the higher Primates may well have been derived (Remane, Werth, Abel, etc.). The incisors are implanted vertically, while the canine is blunt and does not project above the level of the adjacent teeth. The premolars are quite simple, the deutero-cone in the last premolar being represented by a slight basal thickening of the cingulum. In the molars the paraconid has disappeared, but the outer and inner cusps show a primitive character in that they alternate slightly with each other instead of lying in the same transverse planes as they do in the typical anthropoid molar. The talonid lies on a level with the trigonid, and the contrast between the anterior and posterior moieties of the primitive eutherian tuberculo-sectorial tooth is no longer to be seen (Fig. 24, E and Fig. 27).

Parapithecus may most aptly be compared with the modern gibbon, but in the latter various specializations and progressive modifications are to be noted such as the enlarged and tusk-like canines, the two sub-equal cusps of the last premolar, and the transversely arranged pairs of cusps in the molars. On the other hand, the dentition of *Parapithecus* shows some similarities with that of the tarsioids, and it may, therefore, possibly be regarded as a structural link between the Tarsioidea and the Anthropomorpha.

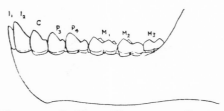

FIG. 27.—LATERAL VIEW OF THE LOWER TEETH OF *Parapithecus* × 2. (After Schlosser).

The outstanding morphological characteristics of the dentition of the Anthropoidea are, in general, spatulate incisors, bicuspid premolars, quadritubercular upper molars with a cingulum hypocone, and quadritubercular lower molars with the loss of the paraconid. But, as we have seen, these progressive characters are not fully developed in some of the known primitive monkeys. It may be inferred, indeed, from a study of such forms as *Parapithecus*, *Apidium* and *Callicebus*, that the ancestral type from which the Anthropoidea were derived would have possessed characters at least as primitive as the following :

Dental formula $\dfrac{2 \cdot 1 \cdot 3 \cdot 3}{2 \cdot 1 \cdot 3 \cdot 3}$.

Incisors slightly spatulate and not procumbent.
Canines but slightly projecting.
Premolars practically unicuspidate, with certainly no more than a small orimentary deuterocone.
Upper molars quadritubercular, with a cingulum hypocone.
Lower molars still retaining the paraconid (at least on the first molar), with a raised talonid and well-developed hypoconulid.

It is of the highest importance to note that in the progressive tarsioids of Eocene times the dentition approaches in many respects that of the Anthropoidea. Thus, as recorded in the previous section, in *Necrolemur* and *Microchœrus* the upper molars had acquired the quadritubercular pattern, while in the lower molars the paraconid is rudimentary or absent, and the talonid is raised up to the level of the trigonid. In some extinct tarsioids, also, the premolars were incipiently bicuspid, and in this sub-order they avoided the molarization which is evidently a definite evolutionary tendency in the lemuroids. In certain of the Eocene Tarsioidea the dental formula was identical with that of the Platyrrhines (*e.g.*, *Hemiacodon Washakius*); in others it is the same as the Catarrhines (*e.g.*, *Anaptomorphus*). The question arises as to the possible polyphyletic origin of the monkeys, for it has been suggested that perhaps the Platyrrhines were derived from American tarsioids and the Catarrhines from European tarsioids. So far as the macroscopic structure of the teeth is concerned, however (and ignoring for the moment evidence from other sources), there is no adequate reason for maintaining such an independent origin of the two groups. That is to say, both Old and New World monkeys may well have arisen from a common ancestor which had already acquired a pithecoid status in respect of its dentition. Thus it may be assumed at this juncture that the early monkeys had the Platyrrhine formula with three premolars, and that with the derivation from such a group of the Catarrhines, the most anterior premolar (P2) was lost.

In the subsequent evolutionary history of the Anthropoidea various specializations emerged such as the hypertrophy of the canines and the bilophodonty of the molars, but in the primitive Anthropomorpha the dentition retained a generalized character which has been transmitted with remarkably little change to Man himself. If *Parapithecus* really comes into the category of the Anthropomorpha (as Werth believes), it is quite reasonable to suggest that the anthropoid apes have themselves been derived directly from a tarsioid ancestry—without the intervention of a pithecoid stage such as might be represented by the primitive cercopithecid *Apidium*, in which the molars were not yet distinctly bilophodont. It might even be argued (on the basis of the dentition) that the cercopithecids are really specialized descendants of a primitive anthropomorph. It will be assumed for the moment, however, that the Cercopithecidæ and the anthropomorph apes are the common derivatives of a primitive form which, in its general structural features, would legitimately be regarded as a generalized type of Catarrhine monkey.

In discussing the dentition of the Tarsioidea, it was seen that the Lemuroidea (as they are at present defined) do not come into consideration as an ancestral group. The same arguments apply to the question of a lemuroid ancestry for the Anthropoidea. The dentition of modern lemurs is far too specialized to have provided a basis for the evolutionary development of the monkeys, and the known fossil lemuroids also manifest divergent modifications, though of less degree. This is especially the case with the Notharctinæ, because of the aberrant nature of the fourth cusp in the quadritubercular upper molars of this group. It may be noted that Gregory has suggested that in *Callicebus* the hypocones are " apparently pseudohypocones " like those of *Notharctus*. But this suggestion rests on no convincing evidence, and, on the other hand, it is probable that the hypocones of other Platyrrhines, as well as those of the Old World Primates, are true cingulum hypocones. Thus there seems here to be a rather serious objection to the hypothesis that the Notharctinæ represent the ancestral group of Primates from which the Anthropoidea have been evolved.

In a general summary of the evidence of the dentition in the Primates we may perhaps draw the following inferences. The variations in the structural pattern of the teeth in the Primates are clearly of the greatest value in the elucidation of natural affinities. No doubt there is on the whole a relation between function (e.g., nature of diet) and tooth structure, but it hardly seems that this conception will account for the more minute differences. Thus it has never been in any way shown that the variations in the diet of the different lemurs and *Tarsius* are of such an order that they actually determine the development in one member of prominent hypocones, in another of mesostyles, in a third of molarized premolars, and so forth. In other words, these small differences have not been proved to have any survival value. The same applies—as already indicated—to the curious specialization of the front teeth which has evolved independently in the Lemuriformes and the Lorisiformes. Consequently, these anatomical differences in the crowns of the teeth provide strong morphological evidence for assessing real relationship, and similarities are evidently not always to be explained away as the results of convergent evolution dependent upon similar necessities of life. Clearly the Primate dentition was derived from a very generalized eutherian type of which the formula was $\frac{3 \cdot 1 \cdot 4 \cdot 3}{3 \cdot 1 \cdot 4 \cdot 3}$. The incisors were not procumbent, styliform, or spatulate, the canines did not project more than a slight degree, the premolars were all simple with a well-developed paracone but with little more than a suggestion of a deutero-cone rising from the internal cingulum, the upper molars were of a simple tritubercular type, and the lower molars tuberculo-sectorial. From a mammalian group with a generalized dentition of this type, the Lemuroidea and Tarsioidea must presumably have diverged in the very earliest stages of Primate evolution. In the Eocene representatives of the Lemuroidea the dentition was still of a primitive pattern, showing, however, the loss of one incisor and a definite tendency towards molarization of the last premolar and the development of a conspicuous hypocone (or pseudohypocone) in the upper molars. Some of the Tarsioidea retained the full complement of teeth of the primitive placental dentition (at least in the lower Eocene), others lost incisors, and, as a group, they show a characteristic tendency to early reduction of the premolar series and (in American genera) to a peculiar specialization of the last premolars. The tarsioids, however, do show many progressive variations towards the dentition characteristic of the Anthropoidea, and, as far as the dentition alone is concerned, the latter may well have been derived from a form similar to *Necrolemur* or *Microchœrus*, but in which certain specializations seen in these two genera had not yet been attained.

The Microscopical Structure of Primate Teeth

In 1922 Thornton Carter described the histology of the enamel in the teeth of a number of recent and extinct Primates.[11] His observations were made on two features: (1) the degree of penetration of the enamel by "dentinal tubules," and (2) the pattern of the enamel prisms. With regard to the first character, there is very considerable variation even within the limits of the Lemuroidea, the penetration being very slight in Lemuridæ, rich in the Indrisidæ and Lorisidæ (except *Nycticebus*), and almost absent in *Chiromys*, the Plesiadapidæ and the Notharctinæ. The enamel pattern of the Lemuriformes (including *Notharctus* and *Pelycodus*) is identical with that of the Catarrhines, while the pattern in the Lorisiformes is similar to that of the Tarsioidea and Platyrrhines. This observation has led to the suggestion (Tate Regan[7]) that the Platyrrhines and Tarsioidea are derived phylogenetically from the lorisiform group, while the Lemuriformes represent the ancestral basis of the Catarrhines. Such a conclusion, however, is entirely negatived by the abundant morphological evidence of a contrary nature derived from other anatomical systems, and is, moreover, not in accord with the palæontological evidence at present available. It is reasonable, in the face of all the data at hand, to accept Thornton Carter's own implication that, in the contrasting types of enamel pattern, " the Lemuroidea . . . parallel the Anthropoidea."

References

1. ABEL, O. : Die Stellung des Menschen im Rahmen der Wirbelthiere. Jena, 1931.
2. GREGORY, W. K. : On the Structure and Relations of Notharctus. Mem. Amer. Mus. Nat. Hist., vol. iii., 1920.
3. GREGORY, W. K. : The Origin and Evolution of the Human Dentition. Baltimore, 1922.
4. OSBORN, H. F. : Evolution of Mammalian Molar Teeth. New York, 1907.
5. SIMPSON, G. G. : A Catalogue of the Mesozoic Mammalia. London, 1928.
6. STEHLIN, H. G. : Die Säugethiere des Schweizerischen Eozäns. Abhand. der Schweiz. Paläont. Gesellsch., vols. xxxviii. and xli., 1912 and 1916.
7. TATE REGAN, C. : Ann. and Mag. Nat. Hist., vol. vi., 1930.
8. TEILHARD DE CHARDIN, P. : Sur Quelques Primates des Phosphorites du Quercy. Ann. de Paléont., vol. x., 1916.
9. TEILHARD DE CHARDIN, P. : Les Mammifères de l'Éocène inférieur français. Ann. de Paléont., vols. x. and xi., 1916.
10. TEILHARD DE CHARDIN, P. : Les Mammifères de l'Éocène inférieur de la Belgique. Mem. du Musée Royal d'Hist. Nat. de Belgique, 1927.
11. THORNTON, CARTER, J. : On the Structure of the Enamel in the Primates. Proc. Zool. Soc., 1922.
12. WERTH, E. : Parapithecus, ein primitiver Menschenaffe. Sitzungsber. d. Gesellsch. naturforsch., 1918.

13. WORTMANN, J. L. : Evolution of Molar Cusps in Mammals.   Amer. Journ. of Phys. Anthrop., vol. iv., 1921.

14. Numerous separate reports on fossil Primates by Matthew, Granger, Gidley and Simpson, mostly in the publications of the American Museum of Natural History.

CHAPTER V

THE EVIDENCE OF THE LIMBS

THE fundamental structure of the limbs in Primates is remarkably primitive.   This fact has been brought out very clearly by Wood Jones in his book " Arboreal Man."   Considering all the evidence derived from a study of generalized mammalia and their reptilian forerunners, there is good reason to assume that in the Tetrapoda the limbs initially were capable of some degree of grasping or clinging power as well as providing a locomotor apparatus.   These functions, moreover, were associated anatomically with a complete separation of the radius and ulna between which some degree of rotation (pronation and supination) was possible, with a relatively freely movable shoulder joint and a well-developed clavicle, with a separate tibia and fibula, and with pentadactyl extremities in which the individual digits could be well abducted or spread apart.   In a large proportion of mammals, however, which have adopted during the course of evolution a wholly terrestrial habitat with a quadrupedal gait, the body is lifted completely off the ground (in contrast to the usual posture of reptiles), and the limbs come therefore to act as props or supports.   This leads to modifications which aim at increasing the stability of the limbs, even at the sacrifice of some of their primitive mobility.   The limbs in this case, therefore, serve mainly as a means of support and progression and lose to a greater or lesser degree their prehensile functions.   Structurally, this change of function is manifested in a limitation of movement at the scapulo-humeral joint, a tendency to atrophy and disappearance of the clavicle, a fusion of the radius and ulna with consequent loss of a rotating movement between these bones, a fusion of the tibia and fibula, and an atrophy or loss of one or more digits in anterior and posterior extremities.

These structural adaptations to a fully quadrupedal mode of progression have been almost entirely avoided by the Primates, and this appears to be due to the fact that the Primates, from the time when they first began to differentiate from the mammalian group which gave rise to the Eutheria, have maintained an arboreal mode of life. Thus, in this Order, the fore-limbs, and to a considerable extent the hind-limbs, have preserved an ancient simplicity of structure and function. The power of grasping and clinging, which evidently characterized the limbs of the earliest forerunners of mammals, has not only been largely maintained, but has been increased by a greater range of movement at the shoulder joint, by a refinement of the rotatory movement between the radius and ulna, and by the development of friction pads on the digits.

In all recent Primates one or more of the digits is covered on the terminal phalanx by a flattened nail (tegula) instead of a sharp projecting and curved claw (falcula). Falculæ are commonly characteristic of primitive mammals as well as of reptiles, and there is thus good reason for supposing that the Primates have been derived from unguiculate or clawed ancestors. Indeed, within the limits of the Primates all stages are represented between a typical claw and a flat nail. A claw is a specialized epidermal appendage developed for purposes of attack and defence, scratching, scraping, digging and climbing. In its typical form it is laterally compressed and curved, and is closely moulded on the terminal phalanx which itself becomes claw-shaped. A typical flat nail, on the other hand, must certainly be regarded in itself as a degenerative formation, a retrogression from the more elaborate structure of the claw. It is usually curved laterally to a moderate degree, and, while it may be used for scratching purposes, it is otherwise little more than a mechanical support for the digital pad of the terminal phalanx. In some Primates the nail becomes so small and insignificant that it can hardly be regarded as more than a vestigial structure—and it may disappear altogether.

The development of a flattened nail is associated with the increasing importance of the terminal digital pad, which not only provides a more efficient grasping mechanism for animals who find it necessary to indulge in arboreal acrobatics, but also comes to be richly supplied by sensory nerves and to form a tactile organ of a high degree of sensitivity. By means of these tactile organs represented in the finger-pulps, the animal is enabled to explore and scrutinize objects of its environment in a detailed manner and thus (as Elliot Smith[3] has pointed out) to cultivate its faculties of perception and judgment. But, although this elaboration of the digital pads is clearly of a progressive nature, the associated transformation of the specialized claw into a flattened nail is no less a degenerative change. It is important to realize this fact, for some authorities have argued that the well-developed and sharp pointed claws in certain Primates are to be regarded as retrogressions from a stage of evolution in which only flattened nails were present, a complete reversal of what is probably the real line of evolutionary progress. This point will be referred to again in connection with the different subdivisions of the Primates, but it may be noted that it is precisely in those forms in which the digital pads have attained a most conspicuous development that the nails show a marked degree of degeneration, e.g. *Tarsius*, for in this form they are on most digits reduced to minute horny plaques to which it would be difficult to assign a function. It is also important to recall the fact that in some marsupials similar changes may occur. Thus in the little *Tarsipes* the digits of the manus are provided with small, flat, scale-like nails. Lastly, the flattening of the nails is always correlated with a flattening of the terminal phalanges, which adopt a spatulate form and support the terminal pads of the digits.

Lemuroidea

The lemurs are all entirely arboreal, either moving rapidly by running and leaping from branch to branch as in the genera *Lemur* and *Propithecus*, or moving slowly with great deliberation by means of a crawling gait as in *Nycticebus*. In either case both extremities are used for grasping the branches firmly, and the digits are modified accordingly. The fore-limbs are always shorter than the hind-limbs—a proportion which is characteristic of the generalized type of mammal. In the Lorisiformes, however, they closely approximate in size.

In noting the characters of the long bones of the upper extremity, we may refer to Gregory's comparative studies in his monograph on *Notharctus*,[4] an Eocene genus in which these bones present more primitive mammalian traits than in the recent Lemuroidea generally.

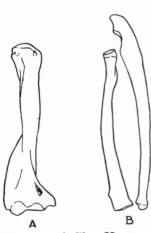

Fig. 28.—A, The Humerus and B, the Radius and Ulna of *Notharctus osborni* × ⅔.

(From a photograph by W. K. Gregory.)

The lemuroid clavicle has a slight curvature, and the scapula in its proportions is similar to that of primitive mammals—*i.e.*, the vertebral border is relatively short. The humerus in recent lemurs has a straight and slender shaft, and, with the exception of *Arctocebus*, the medial supracondylar ridge is pierced by an entepicondylar foramen transmitting the brachial artery and the median nerve. In *Notharctus* the humerus approaches more the generalized mammalian type in its relatively short and somewhat curved shaft, its rather small articular head and flattened trochlear surface, and its prominent supinator ridge (Fig. 28). The latter is evidently associated with a powerful brachio-radialis muscle, which in primitive mammals extends its origin further up the humeral shaft than in Primates generally. It appears that the less extensive origin of the muscle is associated with the capacity for obtaining a complete extension of the elbow, which is characteristic of the brachiating forms of higher Primates.

The humerus of *Notharctus* shows definite Primate features in that the medial epicondyle is rather small (associated with a less strong development of the flexors of the wrist and fingers than is usually the case with primitive clawed mammals), the trochlear surface is separated from the capitellum by a low lip (whereas in most primitive mammals the capitellum and trochlea form a continuous articular surface), and the capitellum is globular (allowing a free rotatory movement of the head of the radius in pronation and supination). Gregory has given a description of the humerus of *Plesiadapis tricuspidens** (a Palæocene lemuroid), and he notes that it is closely similar to the humerus of the primitive pen-tailed tree-shrew (*Ptilocercus*), the supinator crest being more distinct and the shaft relatively shorter in the former.

The radius and ulna in the recent Lemuroidea are always separate, rather slender, and fairly straight bones. In *Notharctus* the shafts of these bones are flattened and show a marked curvature.

The hand exhibits some rather aberrant modifications in the Lemuroidea. In some forms, *e.g. Hemigalago*, the palmar pads show the generalized mammalian pattern—that is to say, there are two proximal (thenar and hypothenar) pads and four distal or interdigital pads, all of which preserve their individuality. In other forms, such as *Lemur*, there is some degree of fusion between adjacent pads—*e.g.*, between the first and second interdigital, the first interdigital and thenar, or the fourth interdigital and hypothenar pads, or, lastly, the hypothenar pad may show a tendency to split into two parts. In some lemuroids, as *Nycticebus*, the individual pads are

* *Nothodectes gidley.*

merged with each other to such an extent that their outlines
are but feebly indicated, and in this feature an approximation
is shown to the higher Primates.

As regards the digits, a characteristic of all lemurs, recent
and extinct (at least in so far as the manus is known), is the
relative elongation of the fourth digit. It has been pointed
out (Morton) that such a disproportion in the digital formula
may be related to the fact that when the hand grasps a bough
of relatively large diameter, a more secure hold is obtained if
the palm is placed obliquely across it with the thumb opposed
to and meeting the outer digits round the bough, in this way
increasing the span of the grasping hand.[5] Thus the
lemurs—which are relatively small animals in relation to
the thickness of the branches among which they move—tend
to develop a specialization of the first and fourth digits which
act somewhat as the two blades of a pair of forceps. While the
long fourth digit of the lemurs is perhaps to be " explained "
in terms of such a functional adaptation, it may be questioned
whether the particular environmental conditions of the
Lemuroidea really do determine or demand such a structural
modification. For in the larger lemurs the digital formula
is typically lemuroid, while in the small tarsier, which has a
similar arboreal habitat, the digital formula in the manus is of
the generalized mammalian type—i.e., the third digit is the
longest. It would seem, therefore, that the elongation of the
fourth digit is the expression of a lemuroid trend of development which is not necessarily entirely secondary to functional
requirements, and in which a striking contrast to the other
Primates is afforded. Moreover, it is very important to note
that this is more than a question of a mere proportional
difference in length of the digits, for in all lemurs the small
interossei muscles of the manus and pes are disposed about
the fourth digit as their functional axis,[6] whereas in the other
Primates generally (as in primitive mammals) the third digit
forms this axis. In some lorisiform lemurs the specialization
of the first and fourth digits is associated with retrogressive

changes in the intervening digits (Fig. 29, A). Thus, in *Perodicticus* and *Arctocebus*, the index finger is reduced to a small nail-less stump containing only two vestigial phalanges, while the third digit is considerably shortened. That the lemuroid digital formula is of very ancient origin is shown

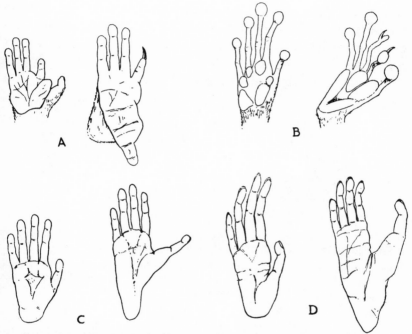

FIG. 29.—THE MANUS AND PES OF A, *Nycticebus*; B, *Tarsius*; C, *Macacus*; AND D, Gibbon.

by the fact that it was already established in *Notharctus* during Eocene times.

In all lemurs except the Aye-aye the digits of the manus are provided with flattened nails instead of claws. In *Chiromys*, on the contrary, all the fingers are sharply clawed. It has been suggested that these claws are really curved and laterally compressed " nails " (tegulæ), because they show a narrow groove on their under aspect. There is no evidence, however, that they have been secondarily derived from flattened nails, and it seems improbable (as shown above) that such a change should occur. In so far as the claws of *Chiromys* are marked by a palmar groove, they are less specialized than those of many quadrupedal mammals and evince a very slight tendency toward the open nail of higher Primates, but they are none the less claws. The retention of claws on all the fingers of the Aye-aye is quite in harmony with the fact that this aberrant lemur retains in many other anatomical features persistent evidence of its derivation from a mammalian group of a very primitive eutherian type. It is interesting to note that in *Notharctus* the only available ungual phalanx indicates that the

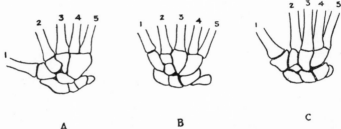

FIG. 30.—THE CARPUS OF A, *Lemur*; B, *Tarsius*; AND C, *Macacus*.

The numbers refer to the digits. Note the large size of the unciform, the reduced os magnum and the displaced os centrale in *Lemur*. These lemuroid specializations have been avoided by *Tarsius* and the Anthropoidea.

nail of this Eocene genus was clearly longer and more laterally compressed than in the Lemuridæ, Lorisidæ and Indrisidæ, though not to such a degree as in *Chiromys*. This provides confirmatory palæontological evidence of the derivation of the flattened nail of Primates from a narrow claw.

In the carpus of lemurs several modifications from the primitive carpal pattern are to be noted which apparently are related to the characteristic divergence and stoutness of the thumb in these animals. In the proximal row of bones the scaphoid is wide, but the lunate very small and narrow. In the distal row the os magnum is narrowed by a lateral compression, while the unciform is typically large. The shrinkage of the os magnum may be such that it does not reach up to gain contact with the lunate bone—thus losing an articulation which is characteristic of the generalized mammalian carpus (Fig. 30, A). There is an os centrale which is pushed to the medial side to such an extent that it establishes contact with the unciform instead of being separated from this bone by the os magnum as in the generalized condition (and in higher Primates). Gregory has suggested that these specializations of the carpus may be secondarily related to the extreme divergence

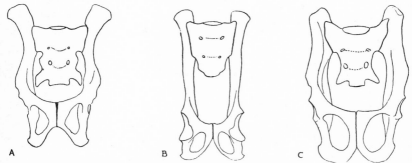

FIG. 31.—THE PELVIS OF A, *Lemur* × ½ (from a photograph by W. K. Gregory); B, *Tarsius* × 1½; AND C, *Hapale* (after Beattie).

of the pollex, which has led to a crowding of the carpal elements to the ulnar side.

Notharctus preserves a more primitive carpal pattern than modern lemurs, for in this fossil the os magnum is relatively larger, the lunate less reduced, and the os centrale probably had " little if any contact " with the unciform. Since the skull structure of *Notharctus* indicates that it belongs to the lemuriform group, it is evident that the lemuriform and lorisiform stocks had separated at an earlier stage, before, in fact, the peculiar carpal specializations of the Lemuroidea had become completely manifested. In other words, these specializations must presumably have progressively developed in the two groups independently.

The pelvic girdle in the Lemuroidea differs but little from that of primitive mammals generally. The ilium is elongated (especially in *Chiromys*) and only slightly flattened to form a blade, while its external surface is usually hollowed out for the attachment of the gluteal musculature (in *Nycticebus* this surface is flat). The iliac surface is narrow and faces ventrally or, in some cases, even slightly laterally (Fig. 36). It articulates only with the first sacral vertebra in *Lemur* and *Notharctus*, and with two vertebræ in *Nycticebus*, *Propithecus*, and *Indris*.[7] In *Microcebus* the ilium is extremely primitive in its rod-like shape and its prismatic form in section. The lemuroid pelvis is narrow, and, associated with this, the crests of ilia diverge rather conspicuously where they serve anteriorly to give attachment to the abdominal muscles (Fig. 31, A). The ischial tuberosities are not markedly divergent as they are in monkeys, correlated with the fact that lemurs are not much accustomed to sitting upright on their haunches. The symphysis pubis is rather short.

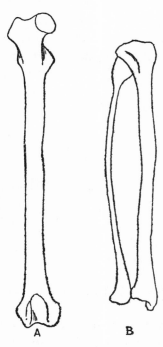

A   B

FIG. 32.—A, THE FEMUR AND B, THE TIBIA AND FIBULA OF *Notharctus osborni* × ⅔. From a photograph by W. K. Gregory.

The femur is slender and has a cylindrical shaft, the greater and lesser trochanters are well developed, and a moderate third trochanter is also present. The tibia and fibula are separate and show a slight bowing, which is considerably accentuated in *Notharctus*. In this form, also, the bones are stouter.

In the lemuroid foot the enlarged and widely abducted big toe, accompanied by a specialization of the outer digits, is a characteristic feature. The fourth digit is always the longest (as in the manus), and the second and third toes may undergo degenerative changes (*e.g.*, in the Pottos). All the toes of the Lemuroidea have flattened nails on their terminal phalanges with the exception of the second, which, being less specialized and important functionally for grasping, retains the more primitive sharp claw. This claw may be used for toilet purposes, and the digit is therefore sometimes referred to as a " toilet digit." In the Aye-aye all the pedal digits are clawed except the hallux, which has a flattened nail. As in the manus of this lemur, the retention of claws is almost certainly a primitive feature.

The plantar pads show a primitive arrangement in some lemuroids such as *Hemigalago*, in which the two proximal and the four interdigital pads are separate. In most genera, however, there is a varying degree of fusion between individual elements, especially between the thenar and first interdigital pads, while in *Nycticebus* the outlines of all the pads are only faintly indicated (Fig. 29, A).

The structural characteristics and variations of the foot skeleton in the Primates have been detailed in a most illuminating monograph by Morton. The results of his studies have been freely made use of in the following discussion. The Primates may broadly be divided into two groups in regard to the mechanism of the foot.[5] In the more thoroughly arboreal types the foot is used to maintain a clinging grasp of the branches, and this leads to a specialization and a wide abduction of the hallux which can be opposed to the outer digits. Thus the foot is used in the manner of a pair of forceps, the blades being represented on the one side by the strong hallux and on the other mainly by the fourth and fifth digits. An animal which uses its foot in such a way will in walking or leaping movements use the anterior tarsal segment (*i.e.*, that part of the tarsus in front of the trochlear surface of the astragalus) as a fulcrum. In the other group—which comprises most of the Primates—the gait is more cursorial, the hallux is not used as an independent digit for purposes of locomotion, and the fulcrum of the foot is represented by the heads of the metatarsal bones.   The structural modifications which are associated with these different uses of the foot are important.

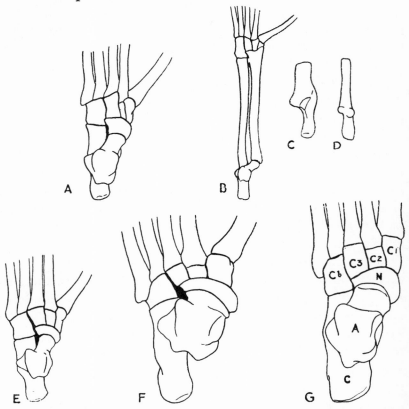

FIG. 33.—THE TARSUS OF A, *Lemur*; B, *Tarsius*; E, *Macacus*; F, *Gorilla*; AND G, MAN (all after D. J. Morton, *Amer. Journ. Phys. Anthr.*, 1924); C, OS CALCIS OF *Omomys* × 3 (after Teilhard de Chardin); D, OS CALCIS OF *Necrolemur* × 1·2 (after Schlosser).

The individual tarsal bones are indicated in figure G: C, Os calcis; A, Astragalus; N, Navicular; C^1, Entocuneiform; C^2, Mesocuneiform; C^3, Ectocuneiform; *Cb*, Cuboid.

All the Lemuroidea use the foot to a greater or lesser degree in maintaining a clinging or perching grasp, and thus belong to the " tarsi-fulcrumating " group of Morton (Fig. 33, A). This specialization is most marked in *Galago*, in which the big toe is widely abducted and the first metatarsal provided with a prominent peroneal tubercle. *Galago* adopts a springing gait, and in correlation with this shows a great elongation of the anterior tarsal segment (os calcis and navicular). This is evidenced in the tarsal pattern by the pushing forwards of the calcaneo-cuboid joint well beyond the level of the astragalo-navicular joint. In *Loris* the most primitive type of tarsus and metatarsus among the Lemuroidea is preserved, and the os calcis is quite short, for this animal moves by a slow crawl and does not require much leverage in the heel. In the Lemuri-formes there is a partial transition to the metatarsi-fulcrumating mechanism, the metatarsals being somewhat lengthened, but the hallux is specialized as in the Lorisiformes, and the length-ening of the os calcis with the forward displacement of the calcaneo-cuboid joint is well marked (thought not to such a degree as in *Galago*). The specialization of the hallux in the Lemuroidea is reflected in the tarsal bones, as is shown by a lateral compression of the mesocuneiform (which is some-what analogous to the compression of the os magnum in the carpus). In *Notharctus* the condition is rather more primitive than in the recent lemuroids : in this genus the metatarsals are shorter and stouter, and the middle cuneiform has not been pressed out of shape.

Thus, in general, we may say that all the Lemuroidea—in their pedal skeleton—show conspicuous evidence of specializ-ation in the " tarsi-fulcrumating " mechanism, a specialization which is reflected in the tarsal and metatarsal pattern even of the less extreme types, such as *Loris*.

Tarsioidea

The mode of progression in the living tarsier is highly specialized, and the fragmentary remains of the limb bones of fossil tarsioids indicate that it had already been adopted to some degree as far back as Eocene times. The animal is entirely arboreal, maintains its hold of the branches by grasping them with both hands and feet, and moves about by leaping from one position to another with great rapidity. The hind-limbs are greatly modified for these saltatory habits, resembling in general proportions those of other jumping mammals (*e.g.*, jerboa, kangaroo), but differing in details of construction. In this connection, *Tarsius* and *Galago* show a striking parallelism.

The fore-limb is much shorter than the hind-limb. The clavicle is slightly **S**-shaped, the scapula has the contour of an isosceles triangle of which the base (vertebral border) is about half the length of the two sides (suprascapular and axillary borders). The humerus is short in proportion to the forearm, with a well-rounded articular head, a strong supinator crest, an entepicondylar foramen, a globular capitellum, and a trochlear surface which is not grooved on its anterior aspect and only faintly demarcated here from the capitellum. The

FIG. 34.—SCAPULA AND HUMERUS OF *Tarsius* (Woollard, *P.Z.S.*, 1925).

radius and ulna are separate and slender, the former being rather weaker than the latter.

In the hand the primitive proportions of the digits are maintained—that is to say, the third digit is the longest (Fig. 29, B). All the terminal phalanges are provided with small flattened and rudimentary nails of a triangular pointed shape, while the terminal digital pads are curiously elaborated to form disc-like expansions covered with papillary ridges by which the animal is enabled to maintain a firm grip of smooth surfaces. The palmar pads are of a generalized pattern, except that the thenar pad is partially fused with the first interdigital pad. In the carpus there is a typical os centrale between the proximal and distal rows of bones, articulating with the lunate, navicular, trapezium, trapezoid and os magnum, but separated by the latter from the unciform (Fig. 30, B). Wood Jones[9] has reported a second os centrale lying on the radial side of the carpus as in *Chiromys* (but this is denied by Woollard[10]). The pisiform articulates with the ulna, the unciform is the largest of the carpal bones, and the os magnum is not laterally compressed. The thumb has undergone no specialization

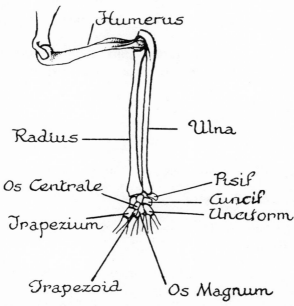

FIG. 35.—THE UPPER LIMB BONES OF *Tarsius* (Woollard, *P.Z.S.*, 1925).

and apparently possesses little or no differentiation of function.

In the pelvic girdle the dorsum ilii is slightly hollowed out on its gluteal surface, and otherwise is elongated in the form of a prismatic rod (Fig. 31, B). The iliac surface faces directly ventrally. The ilium articulates with two sacral vertebræ. The pubic symphysis is quite short and the ischio-pubic arch very slender. The femur is long, slender and straight. There is a third trochanter for the gluteus maximus, and no evident linea aspera. The patella is divided into proximal and distal portions, of which the former is much smaller. The tibia is approximately the same length as the femur, while the fibula is extremely slender and in its distal half completely fused with the tibia. This fusion is a specialized trait in which *Tarsius* stands out among the other living Primates, and in the upper Eocene genus *Necrolemur* a similar degree of fusion between the two bones had already been attained. In this fossil tarsioid, also, the femur was remarkably like that of *Tarsius*, except that the shaft was relatively somewhat stouter.

It is in the foot that the tarsier exhibits the most remarkable specializations. The plantar pads have lost the primitive

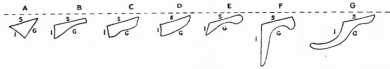

FIG. 36.—TRANSVERSE SECTIONS THROUGH THE ILIUM AT THE LEVEL OF THE LOWER PORTION OF THE ARTICULAR SURFACE OF A, *Macroscelides*; B, *Tupaia*; C, *Galago*; D, *Tarsius*; E, *Cebus*; F, *Hylobates*; G, MAN.

(All after W. L. Strauss, *Amer. Journ. Anat.*, 1929.)

I, Iliac surface; G, Gluteal surface; S, Sacral surface.

mammalian pattern by the fusion of the two proximal pads with each other and the fourth interdigital pad, and of the second with the third interdigital pad. The hallux is large and can be powerfully opposed to the other digits. The fourth digit is the longest, resembling in this feature the Lemuroidea and *Callithrix*. The second and third toes bear sharp curved claws, while the other digits are provided with small flattened and triangular nails. As in the fingers, the terminal digital pads are expanded into discoid formations (Fig. 29, B).

The tarsus is extended to a very unusual degree by the elongation of the os calcis and scaphoid (resembling but surpassing the condition in *Galago*), an extreme modification correlated with the fact that in its foot leverage *Tarsius* has carried

" tarsi-fulcrumation " to an advanced stage in association with saltation (Fig. 33, B). Of the cuneiform bones, the medial is the largest and the middle the smallest. The metatarsal of the hallux is stouter than those of the other digits in relation to the functional differentiation of this digit. The remarkable modification of the tarsal elements in Tarsius is already evident, though to a lesser degree, in the Eocene tarsioids, as shown by the os calcis of *Necrolemur* and *Omomys*. In the former its elongation has reached about two-thirds that of the os calcis of *Tarsius* (Fig. 33, D). To-gether with the evidence of the femur and tibio-fibula of *Necrolemur*, this indicates that the peculiar saltatory mode of progression of the living tarsier was well developed in early Tertiary times. The os calcis in the lower Eocene genus *Omomys* (as figured by Teilhard de Chardin) shows only a comparatively slight lengthening of the fore part of the bone which is not very much more marked than in most of the Lemuridæ and in the primitive Notharctinæ (Fig. 33, C). Nevertheless it indicates definitely the pre-liminary stages which led to the ultimate production of the elongated os calcis of the modern tarsier. Thus it may be inferred that the characteristic tarsioid specialization of the hind-limb was probably initiated at the very beginning of the Eocene, became definitely established during the lower Eocene period, and by the upper Eocene had reached a stage comparable with, though still less advanced than, the final condition represented in *Tarsius*.

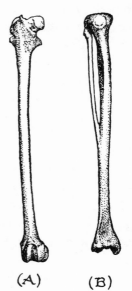

(A) (B)

Femur Tibia & Fibula

FIG. 37.—(A) FEMUR AND (B) TIBIO-FIBULA OF *Tarsius*.

(Woollard, P.Z.S., 1925.)

Anthropoidea

The majority of the Anthropoidea are entirely arboreal. Some, however, have adopted a terrestrial habitat, such as the baboon, which assumes a quadrupedal gait, and Man, a bipedal Primate.   Many monkeys, though typically arboreal, frequently come to the ground and move quite freely on all fours.   The only genera which can in any way be compared with Man in a bipedal mode of progression are the gibbon and gorilla.   The former is fully adapted to an arboreal life, but, besides its power of rapid movement by swinging from branch to branch with its long arms, it is observed also to walk on the ground or along the larger boughs on its hind-limbs in an erect position, holding out its arms in this case in order to balance itself.   There is a difference, however, in the bipedal gait of the gibbon and Man, for while the latter is wholly plantigrade the former does not rest on the heel when it adopts the erect position in the trees.   The gorilla, a large heavy animal, is only slightly arboreal, usually moving about on the ground in the undergrowth of the jungle on all four extremities, but occasionally (as in attack or defence) rearing up and walking on its hind-limbs.

The various modifications in the gait and mode of progression of the Anthropoidea have been dealt with in detail by several authorities (*e.g.*, Abel, Keith, Morton, Gregory), and it will be unnecessary in this treatise to do more than refer to the principal structural changes which are associated with these differences.   The Platyrrhine monkeys and the anthropoid apes (with the exception of the gorilla) are more thoroughly arboreal than the Old World monkeys.   A characteristic feature of extreme arboreal specialization shown in the former groups is a tendency towards lengthening of the fore-limbs (which may be considerably longer than the hind-limbs, as in the spider monkey and the anthropoid apes) and towards atrophy of the thumb.   The latter is associated with the functional specialization of the hand as a " hook " rather than a grasping mechanism, allowing a very rapid release in swinging from branch to branch. In such forms as the gibbon and the orang the feet are not used very much in actual progression among the trees or for support of the body, the arms being employed almost entirely in the movement of swinging from one bough to another. This mode of travel has been aptly termed " brachiation " and associated with it is a disproportionate increase in

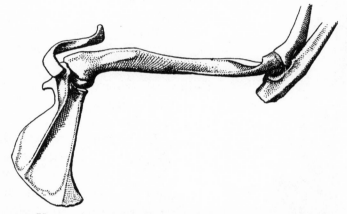

FIG. 38.—UPPER ARM SKELETON OF *Hapale*. (Beattie, *P.Z.S.*, 1927.)

length of all segments of the fore-limb (see Fig. 2). In monkeys which are not such advanced arboreal acrobats, and which run and leap among the larger boughs, the more primitive proportions of the limbs are maintained—that is to say, the fore-limbs are always shorter than the hind-limbs (see Fig. 3).

The clavicle in monkeys is usually bent into an S-shaped curve and is moderately stout. The scapula has a relatively short vertebral border, especially in the Hapalidæ in which it resembles closely the same bone in *Tarsius*. In the higher Primates, Simiidæ and Man, the scapula loses this primitive shape and becomes much longer in a cranio-caudal direction by a lengthening of the vertebral border. The humerus has a slender straight shaft with a well-rounded articular head, a globular and rather large capitellum, and a sharply defined trochlear surface.   These features are associated with great freedom of movement at the shoulder joint, and with a considerable range of pronation and supination of the forearm which can take place independently of flexion at the elbow joint. The supinator crest is less conspicuous than in primitive lemuroids, though better developed in *Cebus* than in most monkeys. The primitive entepicondylar foramen is absent in all Catarrhines, but among the Platyrrhines still persists in *Cebus*, *Chrysothrix and Callithrix*. The radius and ulna are always separate and have slender shafts which are bowed to a slight degree. The articular surface of the head of the radius is circular.

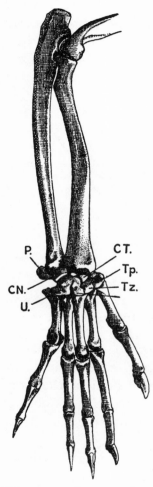

In the manus the primitive digital formula is preserved in all the Anthropoidea, and in many of the Old World monkeys (*Macacus*, *Semnopithecus*) the proportions of the digits approximate closely to those of the human hand (which is indeed a generalized mammalian feature). The diminution of the thumb in certain species of monkeys and apes as an extreme arboreal specialization has been noted above. In the South American spider monkey (*Ateles*) this digit has disappeared altogether except for its metacarpal element.

Fig. 39. — Lower Arm Skeleton of *Hapale*. (Beattie, *P.Z.S.*, 1927.) CN, Cuneiform; CT, Os centrale; P, Pisiform bone; Tp. Trapezium; Tz, Trapezoid; U, Unciform.

In all the Old World Primates (Catarrhine monkeys and anthropomorphous apes) the terminal phalanges are provided with flat thin nails which project but little beyond the finger tips. In many New World monkeys, however (*e.g.*, Cebidæ), the nails are elongated and laterally compressed, forming thus an approximation to a fully developed claw. In the Hapalidæ and *Callimico* the condition is still more primitive (and at the same time more elaborate), for in these animals all the digits of the manus are armed with sharp projecting recurved claws. They are very similar to the typical claws seen in primitive mammals and partly preserved in the lemuroids and tarsioids, but in the presence of a groove on their under surface they show what seems to be an incipient tendency to open out, and thus mark the earliest morphological stage in the transformation of the specialized claw into the less elaborate nail of higher Primates. The retention of claws in the marmoset is of considerable interest in the problem of pithecoid ancestry. In this feature the marmosets contrast rather strongly with other Anthropoidea, though perhaps not so strongly as some authorities seem to think, for, as mentioned above, the Cebidæ provide an intermediate link between the clawed digits of the Hapalidæ and the true nails of the Catarrhines. Attempts have been made to " explain " these claws as the result of retrogressive changes from a true flattened nail. There is no convincing evidence that this is the case, though it may be readily admitted that the marmoset claw is not so highly developed as in some lower mammals. Nevertheless it is almost certainly a persistent primitive structure, and harmonizes with the fact that in this small monkey the skeleton and musculature of the limbs in their arrangement conform very closely to the condition characteristic of primitive and generalized mammals (Beattie).[1] In the embryo *Hapale* (as I have been able to observe through the kindness of my friend Professor J. P. Hill) the claws on all the digits (except the hallux) are somewhat less compressed than those of the adult, resembling more the nails of the Cebidæ. But even at this stage in their development they are not flattened out to form a typical nail such as is seen in the hallux. And, in any case, they retain the propensity for developing sharp and compressed claws—a propensity that has evidently been lost in most other Anthropoidea. The development of claws in

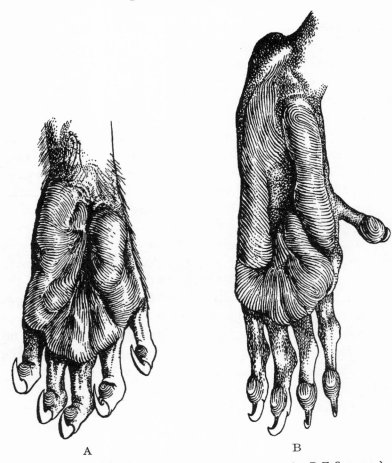

A   B

FIG. 40.—MANUS AND PES OF *Hapale*. (Beattie, *P.Z.S.*, 1927.)

Hapale and *Callimico* is no doubt associated with their arboreal habits, for they can run spirally up and down large tree trunks in the manner of squirrels, which would not be possible if the digits were provided with naked terminal pads as in other monkeys. This, however, does not detract from the morpho-

logical argument that the marmosets have retained the specialized claw which was presumably a feature of the ancestral
Primates. It is more consonant with the facts of the case to
presume that, in adopting its particular mode of life, the
marmoset has avoided the degenerative changes which have
affected the claws in other monkeys for the reason that it
would have been a disadvantage if, in view of its climbing
habits, the terminal digital pads had been elaborated into
tactile organs.

The palmar pads in the Anthropoidea do not show the
definition or distinctness which obtains in lower mammals or
in many lemuroids. In association with the development
of a closer intimacy between the palms and the surface of the
boughs which they grasp, and also correlated with the enhancement of the use of the palms as tactile organs, individual pads
tend to lose their individuality and to merge with each other.
The whole of the palm, also, becomes covered with a rich
system of papillary ridges. In the marmoset, palmar pads
of a typical mammalian pattern are more distinct than in other
monkeys. Thus there are to be recognized thenar and hypothenar pads separated by a wide furrow, and four interdigital
pads of which the first and fourth are blended proximally
with the thenar and hypothenar pads respectively (Fig. 40).
In monkeys generally the thenar pad undergoes atrophy
and disappears.

In the carpus of all the Anthropoidea, with the exception of
the chimpanzee, gorilla, and Man, there is an os centrale. Even
in Man this element is represented in the embryo, but during
development it becomes fused with the scaphoid. While in
the Lemuroidea there is a progressive tendency towards a
diminution of the lunate and a compression of the os magnum,
the Anthropoidea show a specialization in precisely the
opposite direction, for in them the lunate is relatively large
and the os magnum increases in size. Moreover, the os centrale
is kept well away from the unciform by the bulky os magnum
(Fig. 30, C).

In the Anthropoidea generally the thumb has acquired a considerable degree of functional individuality which allows it to be employed with the other digits for grasping purposes. Even in primitive mammalian forms (*e.g.*, the tree-shrew) the hand can be used to some degree as a grasping mechanism, for by a converging flexion of all the digits (which is conditioned by the fact that the carpo-metacarpal joints do not lie in the same plane), together with a movement of adduction, the terminal phalanges are approximated and become to a very slight degree opposed to each other. In the Anthropoidea this movement is greatly enhanced in the case of the pollex by a modification of the carpo-metacarpal joint, the articular surfaces of which, besides being obliquely set in relation to the surface of the palm, become increasingly concavo-convex. Only in Man is the power of opposing the thumb brought to perfection, and in this case, by a combination of adduction, flexion, and a certain degree of rotatory movement at the carpo-metacarpal joint, the palmar surface of the thumb can be pivoted round so as to be brought into direct contact with the palmar surface of the other fingers. In monkeys and apes the thumbs can be incompletely opposed to various degrees, least in the Platyrrhines and probably not at all in the Hapalidæ. It is important to realize that this elaboration of the grasping power of the thumb has its germ in the simple flexion and adduction movements of the pollex in quite primitive mammals, and, indeed, in some marsupials the thumb (and hallux) can be " opposed " as much as in some of the Anthropoidea.

In the smaller Platyrrhine monkeys the pelvis is of a primitive type (Fig. 31, C). It is narrow in the transverse diameter, and the pubic symphysis is rather elongated, sometimes involving the ischial as well as the pubic elements. The ilium is long and narrow, with a very short iliac crest and restricted gluteal fossa, and but little flattening to form an iliac surface ; it articulates practically entirely with the first sacral vertebra. The narrow iliac surface looks almost directly ventrally, and the anterior inferior iliac spine is not very well marked.   In the large Platyrrhines, *e.g., Cebus* and *Alouatta*, the blade of the ilium is slightly expanded, and the iliac surface looks a little medially, while the bone articulates with one and a half or two sacral vertebræ.   In the spider monkey the iliac surface is relatively extensive, and in this case the ilium articulates with three vertebræ.[7]   In all monkeys the ischial tuberosities tend to be splayed out (in contrast to the Lemuroidea) so as to provide a surface for maintaining an upright posture in a sitting position.   In the Old World monkeys the tuberosities are rough and flattened, forming a support for the ischial callosities which are characteristic of this group.   In the larger monkeys and in the Anthropomorpha and Man the ilium becomes progressively expanded to form a flat plate of considerable area.   It is probable that this development of the iliac blade in animals such as the large anthropoid apes, which are accustomed to rapid movement in the trees or to holding their trunk in a vertical position as they sit or climb about, is directly related to body weight.   The expansion provides an increased area of attachment for the gluteal muscles outside and the iliacus muscle inside, but also (probably even more significantly) it allows a great increase in the length of the iliac crest which serves to attach the important muscles of the abdominal wall.   It is often assumed that the expansion of the ilium provides a direct support for the abdominal viscera in the erect or semi-erect position of the body.   Such a misconception seems to have arisen from the idea that the pelvis is to be regarded as a sort of basin.   Even in Man, if the bony pelvis is orientated correctly, it is manifest that the pelvic surface of the ilium faces forwards and inwards, and not upwards to any significant extent.

In the small gibbon the proportions of the narrow ilium resemble closely those of the Catarrhine monkeys, and contrast with the expanded ilium of the large apes.   The conspicuous broadening of the pelvic cavity in Man is partly associated with the adoption of a fully erect posture and bipedal habits, and partly (in the female) with the necessity of accommodating the relatively large fœtal head during parturition.

The femur in the Anthropoidea is characteristically long and slender, and the greater and lesser trochanters are prominent processes. In the Hapalidæ there is also a distinct third

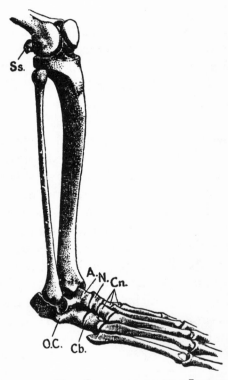

FIG. 41. — FEMUR OF *Hapale*. (Beattie, *P.Z.S.*, 1927.)

FIG. 42.—THE SKELETON OF THE LEG AND FOOT OF *Hapale*. (Beattie, *P.Z.S.*, 1927.) A, Astragalus; N, Navicular; Cn, Cuneiform bones; Cb, Cuboid. O.C., Os calcis.

trochanter as in lower Primates, but this is rudimentary or absent in other Platyrrhines, and absent in the Catarrhines. The tibia and fibula are always separate and have long, slender shafts.

The digital formula of the foot in monkeys and apes is of a

primitive type, the third digit being the longest, with the exception of *Callithrix* in which the fourth digit is slightly longer than the third[7] (thus paralleling the typical lemuroid condition). The toes are all provided with flattened nails except in the case of the Hapalidæ and *Callimico*, in which only the hallux has a true nail while the other digits bear claws of the same type as those of the manus. The big toe is usually large and can be adducted to a marked degree for use in grasping. In Man this power of independent movement has apparently been lost, and in the marmosets it is somewhat limited. But although in the latter the hallux is reduced in size, it is widely divergent from the other digits, it can be adducted and abducted through a considerable range, and " is used in exactly the same way as the hallux of *Cebus*—that is, to obtain a grip on the tree-trunk in addition to the support obtained from the clinging claws " (Beattie).

The plantar pads—as in the palm—merge with each other to a considerable extent. In the marmoset the two proximal pads are distinctly separated from each other, the first inter-digital pad is completely merged in the distal end of the thenar pad, and the three lateral interdigital pads are not sharply individualized (Fig. 40, B).

The characters of the pedal skeleton of monkeys are asso-ciated with the fact that most of these animals adopt to some extent a quadrupedal cursorial gait on the ground or on the larger boughs of the trees in which they live. In using the heads of the metatarsals as the fulcrum in the leverage of the foot they belong to the " metatarsi-fulcrumating " group of Primates, which (in contrast to the tarsi-fulcrumating group) is structurally characterized by a lengthening of the metatar-sals, while the bones of the mid-tarsal region not only pre-serve their primitive pattern but may even show an actual shortening.

In terrestrial Old World monkeys the digits are short relatively to the metatarsals, and the hallux tends to be reduced in size though retaining its power of grasping. In the special-

9

ized arboreal forms, however, such as the gibbon or orang, the digits are greatly increased in length. The grasping power of the hallux is characteristic of all the Anthropoidea with the exception of Man, and, on analysis, is found to be the result of an adduction of the hallux towards the other digits combined with a converging flexion of all the toes. Structurally this functional differentiation is associated with a distinct torsion of the shaft of the first metatarsal bone, besides a characteristic conformation of the articular surface between this bone and the entocuneiform. There is convincing evidence that the human hallux is derived phylogenetically from such a grasping type, for not only does the shaft of the metatarsal bone show the same kind of torsion and the entocuneiform a curved facet (albeit considerably flattened), but the big toe in Man is supplied with all the necessary musculature which is required for producing elaborate movements even though these movements are no longer possible.

The functional axis of the foot leverage in most Anthropoidea corresponds to the line of the third digit, as in the primitive mammalian foot, and this is reflected in the greater length of this digit (with the exception of *Callithrix*). But in the anthropoid apes, as well as in the larger Platyrrhines, this axis is shifted medially to a position between the first and second digits—*i.e.*, to the axis of the grasp. This change is evidenced anatomically by an increase in the length of the second matatarsal, which now becomes the longest of the series. The medial position of the axis is characteristic of the human foot, and this fact, together with the size and stoutness of the hallux in Man, may be taken as a further indication that the human foot is derived from an arboreal type somewhat (but by no means entirely) similar to that of the larger apes.

The tarsal pattern in monkeys is in general more primitive than that of the Lemuroidea, for they have avoided the high specialization and wide abduction of the hallux, the lateral compression of the middle cuneiform, and the elongation of the mid-tarsal region. The Hapalidæ preserve the primitive proportions of the tarsal, metatarsal and phalangeal elements to a remarkable degree. The hallux in this group shows a slight degree of reduction, but is capable of grasping movements as in other monkeys, and—in conformity with this—the first metatarsal bone shows a characteristic torsion, while the terminal phalanx of the big toe bears a flattened nail. The generalized type of tarsus is also but little modified in the terrestrial Catarrhines. In the large anthropoid apes (orang, chimpanzee and gorilla), but not in the smaller gibbon, the anterior tarsal elements (cuneiforms, cuboid and navicular) have undergone a conspicuous shortening. Morton has shown very clearly that this is the direct result of the weight of a large animal on a foot which has already become completely adapted for arboreal life. In such a type of foot the weight of the body is transmitted to the anterior tarsal segment, and this factor is associated with what appears to be a " crushing " or compression of the anterior tarsal elements. The interesting point is that the anterior tarsal elements in the human foot have retained the primitive pattern. Thus it can only be presumed that the early ancestors of the human stock must have forsaken the trees and adopted a terrestrial mode of progression at a stage when they were approximately no larger or heavier than the modern gibbon, and that they increased in bulk *after* they had assumed a truly erect posture— using the heel for the support of the body weight and so largely relieving the anterior tarsal elements from the crushing effect which has distorted them so conspicuously in the large anthropoids. This is a very far-reaching conclusion, for it clearly leads to the inference that many of the strikingly humanoid characters of the large apes which are definitely associated with their size—*e.g.*, the volume and complexity of the brain and the proportions of the skeletal elements—were acquired in this group independently of human evolution. The gorilla has adopted a terrestrial life, but he adopted it too late to acquire anything like the bipedal facility of Man. A much too protracted evolutionary history in the trees has led to arboreal specializations of such a degree that the gorilla foot is quite unable to serve the characteristic functions of the human foot. In its mid-tarsal region the human foot shows more similarities with the primitive condition in the terrestrial and cursorial Catarrhine monkeys than with the large anthropoid apes.

The Evolutionary Origin of the Primate Limbs

We may now summarize the general inferences which can be drawn from the limb structure of the Primates and, in doing so, take the opportunity of emphasizing again certain points which are of considerable significance.

There is much evidence in support of the thesis that all eutherian mammalian Orders were originally derived from a generalized arboreal mammal. There can at least be not the slightest doubt that the Primates are the evolutionary product of a small arboreal precursor not dissimilar in size and general proportions to the modern tree-shrews.

The lemurs are pre-eminently arboreal specialists. In all lemurs, recent and extinct, whose limb skeleton is known, the extremities are clearly adapted for securing a clinging or perching grasp of the branches. In the Eocene *Notharctus* the manus and pes had already acquired a biramous character, the pollex or hallux being used in opposition to the other digits. In the lemuroid manus this specialization has led to some characteristic modifications such as the elongation of the fourth digit, the compression of the os magnum, the reduction of the lunate, and the establishment of an articular contact between the os centrale and the unciform. These features are present or foreshadowed in *Notharctus*. It appears, therefore, that the earliest ancestral stock which gave rise to the Lemuroidea would quite probably have shown these specializations in an incipient form. The retention of claws on all the fingers of *Chiromys* leads to the assumption that this primitive feature was also present in such an ancestral type. These claws may have presented grooves on their under surface, thus on the one hand being less specialized than the claws in many lower mammals, and on the other hand representing the earliest stage in the opening out of the claw to produce ultimately a typical flattened nail. Nevertheless, it would be pedantic to regard such appendages as compressed nails rather than true claws. Judging from the primitive condition of the friction pads in *Hemigalago*, it may also be assumed that the ancestral protolemuroid hand preserved the primitive mammalian pattern of separate and well-defined carpal pads. Such a manus would no doubt have been capable of grasping functions and must have resembled closely that of the tree-shrew. In these small animals the pollex can be considerably abducted from the other digits and, in fact, as is shown in a photograph taken by me of a living tree-shrew (*Tupaia minor*),* may be used in opposition to the rest of the manus for clinging to branches. The characters of the humerus of *Notharctus* and *Plesiadapis* indicate the primitive structure of the upper arm in the earliest lemuroids. There seems no doubt, from a consideration of this evidence, that the fore-limb was from the earliest times capable of free movement at the shoulder joint (in combination with a well-developed clavicle) and at the elbow and the radio-ulnar joints. Probably, also, the fore-limb was rather short in comparison with the hind-limb.

As regards the hind-limb of lemurs, clearly the ancestral lemuroid stock must have possessed a pelvic girdle at least as primitive as that of *Microcebus* or *Galago*. It may be inferred, therefore, that, as in the most primitive mammalia (*e.g.*, the opossum), the ilium was elongated and slender, with a very short crest, a narrow flat gluteal surface facing laterally and a narrow flat iliac surface facing ventrally and slightly laterally. The sacro-iliac articulation involved not more than the first sacral vertebra. In the Tupaiidæ a very close approach is made

* Reproduced as Fig. 10 in Elliot Smith's " Essays on the Evolution of Man."³

to the typical modern lemuroid condition in a slight widening of the dorsum ilii and a hollowing out of the gluteal surface. The pelvic cavity of the ancestral stock was narrow, and the pubic symphysis probably rather short, for Wood Jones has demonstrated that the latter feature is a concomitant of true arboreal life. It is obvious that such a pelvic girdle corresponds to a very generalized and primitive eutherian type. The femur was relatively slender, probably bearing a third trochanter and closely resembling in contour that of *Notharctus* and *Adapis*. The tibia and fibula were quite free, but almost certainly more curved and stoutly built than in modern lemurs.

In the foot a wide abduction and hypertrophy of the hallux must have been very early acquired in association with the adoption of a tarsi-fulcrumating mechanism, and with this specialization the fourth digit became unduly lengthened. Such a biramous type of foot led to changes in the tarsus such as the diminution and compression of the mesocuneiform. In almost all lemurs the power of leaping among the branches is evidenced structurally by some elongation of the anterior tarsal segment, and this is even shown to some degree in *Notharctus* and the still more primitive *Pelycodus* by the extension forwards of the anterior moiety of the os calcis. In *Adapis*, on the other hand, the os calcis is of a more generalized type with a short sustentaculum tali and a short anterior moiety, somewhat resembling in these features the slow-moving Lorisidæ. Thus it may be inferred that the ancestral lemuroid stock was not a very active arboreal creature. In most lemuroid genera, however, an increasing arboreal agility has been accompanied by an elongation of the anterior tarsal segment which is foreshadowed in the Notharctinæ, quite distinct in such forms as *Lemur* and *Propithecus*, and carried to an extreme in the Galagidæ. This specialization of the lemuroid foot (which was thus incipiently shown even in Eocene times) could hardly have led on to the more generalized tarsal pattern found in the Anthropoidea. The persistence of a claw in all lemurs on the second pedal digit (which has, as it were, been passed over by the trend of specialization affecting the hallux and the outer digits) suggests the derivation of the lemuroid foot from a clawed type, and, in fact, in *Chiromys* claws are still borne on all the digits with the exception of the great toe.   It is probable, therefore, that at least the four outer toes of the lemuroid ancestor were clawed.*

Tarsius shows an astonishing contrast in its extremities, for while the hind-limbs are in some ways more divergently modified than in any other Primate, the fore-limbs preserve to a considerable extent a primitive mammalian structure. In the presence of two claws on the pedal digits, also, it is more primitive than most lemuroids.   It has been pointed out that the specialization of the hind-limb for saltatory progression among the branches was already evident in Eocene tarsioids.   It is clear, therefore, that these peculiar modifications were attained at a very early stage in the evolution of the group.   If we consider the problem of their original derivation from the common ancestral Primate stock, it becomes evident that the Tarsioidea almost certainly developed independently of the Lemuroidea and prior to the time at which the latter were to be distinguished as a separate sub-order by the acquisition of their own characteristic modifications.   For in all the modern and extinct lemuroids of which the appendicular skeleton is known to any extent divergent specializations are distinct in the fore-limb of such a nature that it is hardly possible to regard them as representing forms ancestral to the tarsioids.   It follows, therefore, that the

* The structural modifications which might be anticipated in a lemuroid type of foot were it to become adapted for a terrestrial gait are suggested by the example of the marsupials.   The foot in primitive arboreal marsupials is very similar in many respects to that of lemurs, with a divergent hallux, elongation of the fourth digit and reduction of the second and third digits.   In terrestrial forms such as the kangaroo, it is the specialized fourth digit which becomes the main supporting digit. *Dendrolagus* represents a form which has returned once more to arboreal life after a terrestrial phase.   It preserves a foot structure essentially of the kangaroo type and shows no tendency to revert to the more primitive and seemingly more appropriate foot of the original arboreal type.

close resemblances in the hind-limb of *Tarsius* and *Galago* are the result of parallel evolution, for, as we have seen above, the specializations of the hind-limb and foot peculiar to the Galagidæ were undoubtedly not present in the immediate ancestors of the lemuroids. A survey of the limb structure of the lemuroids and tarsioids leads to the inference that a common ancestor which might have given rise to the divergent groups must have possessed limbs similar to those of a generalized arboreal insectivore type. On the other hand, the hind-limb skeleton would possibly betray the Primate affinities of such a form, for the evidence suggests that in the early evolutionary development of the Primates among other mammalian Orders the hind-limb (and especially the foot) acquired distinctive Primate characters in advance of the fore-limb.

Turning now to the Anthropoidea, it has been made evident that in the upper extremity certain of the monkeys have preserved a simplicity of structure comparable with that of *Tarsius*, and in some points even more pronounced. The primitive construction of the arm bones in the Hapalidæ indicates a phylogenetic derivation from quite a generalized eutherian type. Probably the movement at the scapulo-humeral, elbow and radio-ulnar joints was more free in the early forerunners of the monkeys than in the direct ancestors of lower Primates. In the carpus a close resemblance to *Tarsius* is seen, and the trend of evolutionary development here seems to have diverged *ab initio* from the typical lemuroid type, for the lunate and the os magnum became large and the lemuroid hypertrophy of the unciform was avoided.

The manus in the marmosets gives a more suggestive indication of the primitive nature of the ancestral Platyrrhine stock, even though in the former there has evidently occurred some degeneration of the pollex. Thus in the Primate group from which monkeys were directly derived it is highly probable that all the digits of the manus had curved and sharp pointed claws, the third digit was the longest, the palmar pads showed a tendency to merge with each other, the thumb could be strongly adducted and flexed into the palm but not opposed to any significant degree, and the fore-limb as a whole was quite distinctly shorter than the hind-limb. As regards the latter, the pelvic girdle in such an ancestral form must have been at least as primitive as that of the small Platyrrhine monkeys, the femur rather long and with a slender shaft, the tibia and fibula quite separate, and the tarsal pattern of a generalized mammalian type such as is still preserved in the marmoset and even in the quadrupedal monkeys such as the macaque. Moreover, the digital formula of the foot (unlike that of lemurs and *Tarsius*) was of a generalized mammalian type, and the hallux could probably be well abducted and used for grasping. Thus it is very unlikely that the pithecoid foot could have been derived during the course of evolution from the lemuroid type with its peculiar specializations. On the contrary, the evidence indicates rather that the ancestral type common to all the Anthropoidea possessed a limb structure of a truly generalized and arboreal type, in which the fore-limb was somewhat similar to that of *Tarsius* but more primitive in the presence of claws, while the hind-limb might well have been typical of an arboreal insectivore except for a somewhat enhanced grasping capacity of the hallux, which perhaps was also provided with a flattened nail of the Primate type.

When the emerging Anthropoidea had reached the stage of evolution represented by the Hapalidæ, the tactile pads on the terminal phalanges of both extremities became increasingly important, and with this the claws (which in the earliest phases of Primate evolution were probably not highly specialized) opened out and underwent retrogressive changes leading to the flattened nails characteristic of the Catarrhine monkeys and apes. In the Old World monkeys a primitive condition of the limbs is maintained to a large degree, but in the more completely arboreal monkeys, and especially in the brachiating anthropoid apes, certain divergent specializations are to be noted. The arms in these forms increase unduly in length, the fingers become long, and the thumb degenerates. These changes have been regarded as of the highest significance in the problem of Man's immediate ancestry (*e.g.*, by Osborn, Wood Jones and others), for in Man the more primitive proportions of the limbs and digits are preserved. In spite of arguments to the contrary, this line of reasoning is quite sound in principle. It is an incontrovertible fact that in the anthropoid apes (including the gorilla) the thumb is atrophied to a greater or lesser degree, and the inference is bound to follow that Man, with his powerful and well-developed thumb, must almost certainly have been derived from a group of ape-like creatures in which such retrogressive changes had not yet occurred. Thus, so far as it is possible to make any definite statement on evidence which is admittedly incomplete on the palæontological side, it may be reasonably concluded that in the line of human ancestry brachiation was never developed to the degree to which it is adopted by the living anthropomorphs. This is not to say, however (as Wood Jones implies), that Man has not been derived from a type which might still be accurately referred to the Simiidæ, even though its limb proportions would hardly have accorded with those of the known anthropoid apes, for such a type would presumably have shown many structural features in common with this group, not only in its skeletal anatomy, but also in the cerebral, dental, visceral, and other systems.

Emphasis may be again laid on the evidence of foot structure, which very convincingly indicates that the forerunners of the Hominidæ diverged from the forerunners of the large anthropoid apes at a point in evolutionary time when the common ancestor had not exceeded the body size of the gibbon. This may be assumed to be the case in view of Morton's studies, and, the assumption having been made, it clearly follows that to speak of a " gorilloid heritage " when discussing the origin of the human foot would be quite misleading. Gregory has recently laid much stress on the point for point resemblances which the limb structure and skeleton of Man show with those of the gorilla in contrast to the other apes and the monkeys. Such close similarities obtrude themselves upon the notice of every anatomist who makes a study of the higher Primates, and have been well established since the time of Darwin and Huxley. But it is clear that these resemblances do not (as Gregory appears to think) necessarily indicate a close affinity in the sense that Man is directly derived from the ancestral stock of the African apes. It may (and probably does) only indicate that the gorilla and Man have both been derived from a fairly distant common ancestor not larger than the gibbon, and from which common potentialities and tendencies for evolutionary development have been inherited. Since the gorilla and Man have approximated rather closely to each other in body size, in posture, in the use of the arms and the legs and in other functions, it is only to be anticipated that the skeletal structure in both these forms will become correspondingly approximated. Gregory seriously considers the possibility of a diphyletic origin of New and Old World monkeys, even though in their skeleton, brain, viscera, etc., members of these two groups may be astonishingly similar (much more so than either would be to the presumed common tarsioid ancestor of Eocene times). If such a conception is valid, corresponding degrees of resemblance between the giant anthropoids and Man need not preclude the probability of their diphyletic origin from a stage of evolution represented approximately by the gibbon (or even, if the accumulation of evidence warranted it, at an earlier stage).

References

1. BEATTIE, J.: The Anatomy of the Common Marmoset. Proc. Zool. Soc., 1927.
2. CLARK, W. E. LE GROS: The Anatomy of the Pen-tailed Tree-shrew. Proc. Zool. Soc., 1926.
3. ELLIOT SMITH, G.: Essays on the Evolution of Man. London, 1927.
4. GREGORY, W. K.: On the Structure and Relations of Notharctus. Mem. Amer. Mus. Nat. Hist., vol. iii., 1920.

5. Morton, D. J. : Evolution of the Human Foot. Amer. Journ. of Phys. Anthrop., vol. vii., 1924.
6. Nayak, U. V. : A Comparative Study of the Lorisinæ and Galaginæ. Unpublished Thesis.*
7. Strauss, W. L. : Studies on Primate Ilia. Amer. Journ. of Anatomy, vol. xliii., 1929.
8. Wood Jones, F. : Arboreal Man. London, 1916.
9. Wood Jones, F. : Man's Place among the Mammals. London, 1929.
10. Woollard, H. H. : The Anatomy of Tarsius Spectrum. Proc. Zool. Soc., 1925.

---

* I have to thank Dr. Nayak for his kind permission to refer to this excellent and important monograph on the Lorisiformes. It formed the main thesis for which the Ph.D. degree of the University of London was conferred on him in 1933, and has not yet appeared in a published form.

CHAPTER VI

THE EVIDENCE OF THE BRAIN

THE comparative anatomy of the central nervous system is of particular value in the elucidation of phylogenetic origins. By the very nature of its functions, it must be quite exceptional for the higher levels of this system to undergo retrogressive changes during the course of evolution, and the principle of the Irreversibility of Evolution, if it applies to any system, must surely apply in this case. For since a progressive elaboration and differentiation of the brain indicate increased powers of apprehending the nature of external stimuli, a capacity for a wider range of adjustments to any environmental change, and an enhancement of the neural mechanisms for effecting more delicately co-ordinated reactions, it is difficult to believe that when such advantages have been acquired during evolutionary development they can again be dispensed with, unless their disappearance is accompanied by very extreme forms of specialization in other directions. If Natural Selection plays a significant rôle in the direction of evolutionary development, it can hardly fail to be a potent factor in the prevention of retrograde changes in the most important controlling centres of the brain. We may infer, then, in a general way, that the structure of the brain in the earliest Primates must have incorporated the primitive features of the most primitive types of brain in the known Primates, and a similar statement might be made in regard to the three sub-orders of the group.

By way of introduction it is convenient first to note the main features of a primitive and generalized eutherian brain, and in order to emphasize the contrast which such a brain shows with the more highly developed structure of the modern Primates, we may use for our purpose the archaic type which is found to-day in the Madagascan insectivore *Centetes* (Tenrec). In general proportions the brain of *Centetes* (see Fig. 43) resembles quite remarkably those of certain primitive Eocene mammals. The olfactory bulb is very large, projecting well

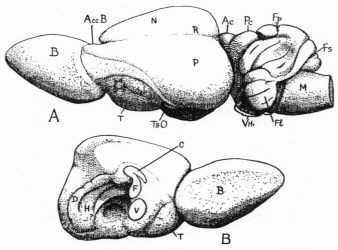

FIG. 43.—A, LATERAL AND B, MEDIAL VIEWS OF THE BRAIN OF AN INSECTIVORE, *Centetes*, TO SHOW THE MAIN PARTS AND THE GENERAL PROPORTIONS OF A GENERALIZED AND PRIMITIVE EUTHERIAN BRAIN × 3. (*P.Z.S.*, 1932.)

Ac, Anterior colliculus; B, Olfactory bulb; C, Corpus callosum; D, Dentate gyrus; F, Fornix commissure; Fl, Floccular lobe of the cerebellum; Fp, Primary fissure of the cerebellum; Fs, Secondary fissure; H, Hippocampus; M, Medulla; N, Neopallium; P, Piriform lobe; Pc, Posterior colliculus; R, Rhinal sulcus; T, Olfactory tubercle; V, Ventral commissure; Vth, Trigeminal nerve.

in advance of the rest of the brain and attached to the latter by a well-defined and broad peduncle. The lateral view shows that a large proportion of the cerebral hemisphere is taken up by the piriform lobe—a cortical structure which forms a part of the rhinopallium (or olfactory cortex) since it receives many of the terminal fibres of the olfactory tract. The dorsal surface of the hemisphere is separated from the piriform lobe by the rhinal sulcus and is formed by a relatively limited and smooth neopallium. The latter is a developmental extension of the cortex which is characteristic of mammalia in contrast to lower vertebrates, and it is concerned with the reception and interpretation of sensory impulses other than those of smell. Thus the progressive elaboration of the neopallium in mammals provides an index of the increasing influence of general sensory, visual and auditory impulses in controlling behavioural reactions. In an animal like *Centetes* it is clear that smell is the dominant sense, but, as Elliot Smith[10] has discussed at length, an arboreal habitat tends inevitably to lead to a reduction of olfactory mechanisms because the sense of smell can hardly play such an important rôle in guiding an animal through its life among the branches as it clearly does with terrestrial mammals in general. On the other hand, arboreal life favours or stimulates the progressive development of tactile, visual and auditory sense organs, with which is associated an elaboration of the corresponding cerebral centres, and this leads to a great enhancement of sensory acuity, affecting all the senses except that of smell. In the higher Primates the neopallium becomes increasingly elaborate and convoluted and the rhinopallium more and more reduced, so that ultimately the latter is completely overshadowed by the former, and the shrunken piriform lobe becomes pushed entirely on to the medial aspect of the hemisphere.

On the base of the brain in *Centetes* there is a large olfactory tubercle, one of the secondary olfactory centres. The medial view of the brain shows a very small corpus callosum (a neopallial commissure) and a big ventral commissure which is largely made up of commissural fibres linking the grey matter of the two olfactory bulbs together. The proportionate size of these two commissures is reminiscent of the more primitive condition pertaining in the marsupials and monotremes, in which the corpus callosum is entirely absent and the ventral commissure conspicuously large. There is also to be seen a plump fornix commissure, which is connected with the hippocampal formation. The latter can be partly seen in Fig. 43B, represented by two strips of cortex—dentate gyrus and hippo-campus—extending downwards from the corpus callosum to the base of the brain. These form another part of the rhino-pallium, developed primarily in relation to the olfactory sense, and are relatively extensive in lowly vertebrates generally. The optic nerves and tracts are of moderate size, their fibres termin-ating mainly in the anterior colliculus in the roof of the mid-brain, and partly in a thalamic nucleus—the lateral geniculate body. The latter is connected by fibre tracts with the occipital pole of the neopallium where the visual cortex is developed. The thalamus is a large sensory centre in the middle of the fore-brain in which sensory tracts end which have ascended from lower levels of the brain. It comprises an important relay station for the projection of sensory impulses on to the cerebral cortex. Besides the lateral geniculate body already mentioned, it contains the medial geniculate body, which receives auditory impulses and is related to the temporal region of the neopallial cortex, and the ventral nucleus, which receives impulses of general sensation (tactile, kinæsthetic, temperature and so forth) and is related to the parietal sensory area of the neopallium. At its dorsal surface is the lateral nucleus (small in primitive mammals), which is a relatively higher functional level of the thalamus not receiving sensory impulses directly from lower levels of the brain. Below the thalamus proper—and reaching to the base of the brain—is the hypothalamus, a region which is closely connected func-tionally and anatomically with the pituitary gland, and which plays an important part in regulating the activities of the autonomic or visceral nervous system.

In the roof of the mid-brain are seen the four corpora quadrigemina or colliculi, of which the anterior pair are related partly to the visual system, while the posterior pair form a mesencephalic auditory centre. The cerebellum is of a simple type in *Centetes*. It has a large median part, the vermis, which is divided by two sulci (the primary and secondary fissures) into anterior, middle and posterior lobes, and a big parafloccular lobe. The latter, together with the vermis, constitutes the most primitive part of the cerebellum. The lateral lobes of the cerebellum are relatively small and simple. They represent a lateral extension of the lobus medius of the vermis, and since they are intimately connected with the neopallial cortex they usually show in the mammalia a degree of elaboration which runs closely parallel with it.

The Primates are characterized by relatively large and elaborate brains, a feature which reaches its highest expression in Man. This increase in the size of the brain involves mainly the cerebral hemispheres and the cerebellum, and is associated with a phenomenal expansion of the neopallium which becomes convoluted in a very complicated manner, with an elaboration of the basal ganglia inside the cerebral hemisphere (especially the thalamus), and with a rich fissuration of the lateral lobes of the cerebellum.

Lemuroidea

The most generally primitive brain among the recent Lemuroidea is that of *Microcebus*,[4] the small mouse-lemur of Madagascar (Fig. 44). In this genus the olfactory bulbs are much reduced in comparison with the brains of lower mammals, and the olfactory tubercles are flattened. The gyrus dentatus and hippocampus are both exposed on the medial surface of the brain from the corpus callosum downwards, while the piriform cortex appears on the lateral surface forming the tip of the temporal lobe of the cerebral hemisphere and demarcated by a shallow but quite distinct rhinal sulcus. The neopallium is relatively simple, showing sulci in two regions only. On the lateral surface is a deep Sylvian fissure, while on the medial aspect of the occipital lobe is a tri-radiate sulcus which is formed of three elements, the sulcus paracalcarinus, sulcus precalcarinus and sulcus retrocalcarinus. Of these, the last is of particular interest, for it is found in the brains of all recent Primates and does not occur elsewhere among mammals except sporadically in the Carnivora. The development of this sulcus is correlated with an expansion of the visual cortex, and it is the result of a folding along the centre of this cortical area.

The Sylvian fissure of the Primate brain is fundamentally composed of two originally separate elements which in lower mammals retain their individuality.[7]   These are the supra-Sylvian sulcus and the pseudo-Sylvian sulcus, of which the former is one of the most primitive sulci of the mammalian brain, while the latter is morphologically rather an inconstant

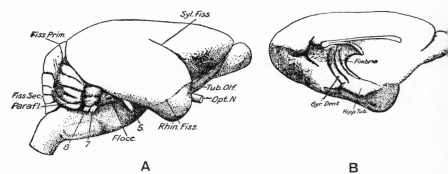

FIG. 44.—A, LATERAL AND B, MEDIAL VIEWS OF THE BRAIN OF THE MOUSE LEMUR, *Microcebus*.   ×2½.   (*P.Z.S.*, 1931.)

formation produced as a kink in the course of the rhinal sulcus (and often loosely called the " Sylvian fissure "). In all living Primates—with the solitary exception of the Aye-aye—the supra-Sylvian sulcus extends down to meet the rhinal sulcus and here blends with the pseudo-Sylvian sulcus so as to form the true Sylvian fissure (Fig. 45, C). In lower mammals* the supra-Sylvian sulcus does not reach the rhinal sulcus in this way, but, on the other hand, it often meets at its upper end the post-Sylvian sulcus (parallel or superior temporal sulcus) to form a wide arcuate fissure (see Fig. 45, B). The corpus callosum in *Microcebus* is long and attenuated, showing slight

* With the exception of the Great Ant-eater (*Myrmecophaga*), which in this respect simulates the Primate brain, as Elliot Smith first showed.

thickenings at the two extremities (genu and splenium), while the fornix commissure is rather reduced.

It is convenient at this point to refer to the fact that the neopallial cortex of mammals is not homogeneous in its structure, but can be mapped out into a number of different areas each of which may be distinguished microscopically. Diagrams showing these cortical areas in *Microcebus* are shown in Fig. 46, and are there compared with a similar chart of the

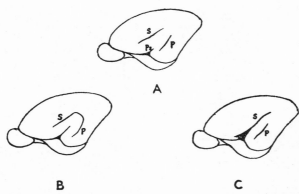

Fig. 45.—Diagrams representing the Formation of the Sylvian Fissure of the Primate Brain, as Elucidated by Elliot Smith.

The generalized condition is seen in A, in which there are three distinct and separate sulci, the supra-Sylvian sulcus S, post-Sylvian sulcus P, and the pseudo-Sylvian sulcus Ps. In many lower mammals and in *Chiromys* the two former sulci become joined to form an arcuate sulcus arching over the pseudo-Sylvian sulcus (diagram B). In Primates generally, the supra-Sylvian merges with the pseudo-Sylvian sulcus to form a true Sylvian fissure (diagram C).

brain of the lesser tree-shrew (*Tupaia minor*). Some of these areas are sensory projection areas where sensory impulses are projected on to the cortex from lower levels of the brain. Thus area 1-3 is concerned with general bodily sensation, area 22 with auditory sensation, and area 17 with visual sensation. This last area is extremely well defined, highly differentiated, and relatively extensive in all Primates, in association with the predominance of the sense of sight in this Order. Among other features it is characterized structurally by an abundance of small granule cells which form two layers, and between these is a conspicuous band or stria of white fibres. This area is usually called the area striata. On the ventral aspect of the frontal pole of the hemisphere there is a small patch of cortex (area 8), which may be called the prefrontal area, and which receives fibres from a part of the thalamus called the nucleus dorso-medialis. This nucleus, further, is connected

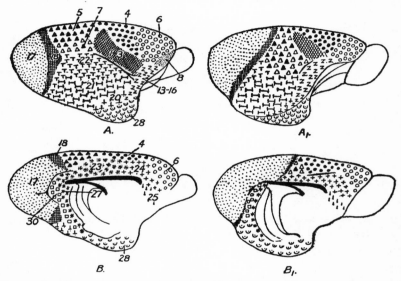

Fig. 46.—Diagrams showing the Different Cortical Areas of the Brain of *Microcebus* (*A.* and *B.*) and the Tree-Shrew, *Tupaia minor* (*A₁.* and *B₁.*) × 2½.

For the significance of the numbers, see the references in the text.
(*P.Z.S.*, 1931.)

with the visceral centres of the hypothalamus. There is evidence which suggests that the prefrontal cortex develops as a mechanism through which visceral reactions (which form an essential element in emotional behaviour and instinctive impulses) may be controlled by the highest functional levels of the brain. It is interesting to note, therefore, that the pre-frontal cortex (and, with it, the dorso-medial nucleus of the thalamus) undergoes a progressive expansion in the Primate series, reaching its highest expression in Man.* Area 4 represents the motor cortex, a motor projection area which gives origin to the pyramidal tracts by which voluntary movements are directly initiated and controlled. The other cortical areas indicated in the diagrams are mostly " association areas." That is to say, they are not concerned so much with the direct reception or emission of impulses from or to lower levels of the central nervous system, but rather with correlating the functions of the projection areas and forming the anatomical substratum of mental processes such as the association of ideas, memory and so forth. As might be expected, these association areas become more extensive in the brains of higher Primates, and, as they expand, they separate more and more the projection areas from each other. In *Microcebus* they are already large in comparison with lower mammals of an equivalent size, especially the parietal association area which separates the general sensory cortex in front from the visual cortex behind. The piriform cortex is represented in the diagram as area 28.

The thalamus in *Microcebus* shows distinctive Primate features in the large size of the lateral nucleus (which leads to a characteristic broadening out of the dorsal surface of the thalamus) and in the elaboration of the lateral geniculate body. The configuration and detailed structure of this important visual centre have provided some significant evidence for the problems of mammalian affinities, and attention may be called to these points here.[5] Transverse sections through the nucleus, showing the arrangement of its constituent cells in different Primates, are represented in Fig. 47. In a simple type of brain in which the visual centres are well developed, such as that of *Tupaia minor*, the lateral geniculate body is situated at the lateral surface of the thalamus (Fig. 47, *A*). The fibres of

* In lower mammals this area is ill defined; my own observations indicate that it develops as a forward extension of the insular region of the cortex (area 13-16), and is not to be regarded as a differentiation of the frontal cortex proper.

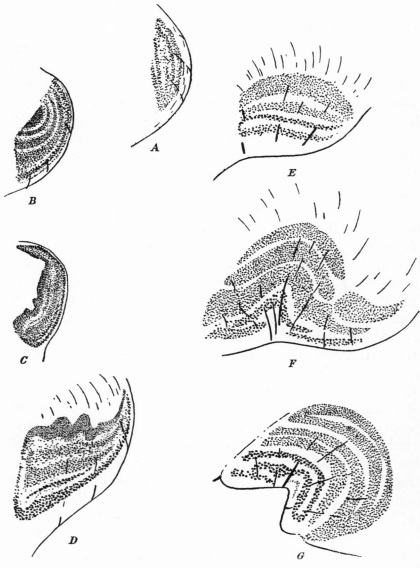

Fig. 47.—The Appearance in Transverse Section of the Lateral Geniculate Body of, A, Tree-Shrew (*Tupaia minor*); B, Mouse Lemur; C, Dwarf Lemur; D, *Lemur catta*; E, Orang Utan (after Kornyey); F, Man; G, *Cercopithecus*, showing the Arrangement of the Cell Laminæ. (*Brit. Journ. Ophthalm.*, 1932.)

the optic tract penetrate it from the lateral convex aspect and from below, while the optic radiations which connect it with the area striata of the cortex leave its medial surface. The cells are disposed in a series of vertical laminæ. In all Primates this lamination is a very conspicuous feature, and as many as six layers of cells may become differentiated. Of these, the two most superficial laminæ are characteristically composed of large cells, and in all the Lemuroidea the most medial laminæ are formed of quite small granular cells. This small-celled element of the lateral geniculate body is definitely to be regarded as a lemuroid specialization. In *Microcebus* these structural features are clearly manifested (Fig. 47, *B*). The laminæ are curved in conformity with the convexity of the lateral surface of the thalamus, the optic tract fibres entering the convex aspect of the laminæ, while the optic radiations are given off from their concave aspect. In larger lemurs (*e.g.*, *Chirogaleus*, *Lemur*) the medial concavity of the laminæ becomes exaggerated, leading to an infolding or inversion of the whole geniculate body (Fig. 47, *C* and *D*). At the same time the most medial lamina is crinkled up by small convolutions which result from and allow of its expansion. With the growth of the lateral nucleus of the thalamus in larger forms, also, the geniculate body becomes displaced more on to the ventral aspect of the brain. Thus, in summary, it may be stated that the lateral geniculate body of the Lemuroidea is characterized (1) by the development of a small-celled element which constitutes the most medial cell lamina, and (2) by the inversion of all the laminæ accompanied by a crinkling of the most medial lamina. It will be seen later that in the Anthropoidea the elaboration of the geniculate body has proceeded in quite a different manner, for in these the laminæ become everted.

The mid-brain in *Microcebus* is proportionately large, and the colliculi form conspicuous elevations. The cerebellum is relatively simple. It has a well-developed vermis and parafloccular lobe, while the lateral lobes are small.

It has been stated above that *Microcebus* presents a brain

that is on the whole the most primitive in all modern lemuroids. It is of interest to refer to the evidence of the conformation of the brain in the extinct genera *Adapis* and
Notharctus. The endocranial cast of *Adapis parisiensis* has
been figured and described by Neumayer.[12] Unfortunately
he mistook the impression of one of the paranasal air sinuses
for the olfactory bulb (as was pointed out by Dr. Tilly Edinger)
and thus represented the brain of *Adapis* as a most extraordinary and quite unique type. By the kindness of my friend
Mr. Hopwood I have been able to examine the endocranial
cavity of a remarkably well-preserved skull of *Adapis parisiensis*
in the British Museum (cat. no. M1345), and by taking measure-

FIG. 48.—THE APPEARANCE OF THE BRAIN OF *Adapis parisiensis* AS
RECONSTRUCTED FROM NEUMAYER'S ENDOCRANIAL CAST AND FROM
A STUDY OF THE INTRACRANIAL CAVITY × 2 APPROXIMATELY.

ments of the cribriform fossa it is clear that the olfactory bulbs
in this species were not very large, each measuring approximately 3 mm. in width, 4 mm. in dorsi-ventral diameter, and
projecting not more than 5 mm. in advance of the frontal pole
of the hemisphere. From my own studies of the skull, and by
reference to the figures of the endocranial cast published by
Neumayer, I have constructed the diagram shown in Fig. 48,
which must represent quite closely the appearance of the
lateral view of the brain of this fossil lemur. It will be
observed that the cerebral hemispheres are not unlike those
of *Microcebus*, showing a powerful development of the temporal lobe and a well-marked sulcus which is evidently a
Sylvian fissure of the lemuroid type. On the other hand,
the occipital lobes are much smaller and are not produced

backwards so as to overlap the hind-brain to the same degree, while the olfactory bulbs were slightly larger (though definitely reduced in comparison with the brains of non-Primate mammals). Viewed from above, the frontal lobes are narrower than those of *Microcebus*, but in the parietal and temporal regions the breadth of the hemispheres in *Adapis* is approximately of the same order. The neopallium was evidently quite smooth on its exposed surface except for the Sylvian fissure. The cerebellum and medulla are relatively large and lie posterior to rather than below the cerebrum. Gregory[11] notes that, judging from fragmentary remains of the endocranial cast, the brain of *Notharctus* was similar to that of *Adapis parisiensis* in its general contour. In both *Adapis* and *Notharctus* the size of the brain as a whole relatively to the size of the body was certainly smaller than is the case in modern Lemuroidea.

Thus in Eocene times the Lemuroidea had already acquired a brain which was definitely of a lemuroid type, showing an advance on the brain of lower mammals in the reduction of the olfactory bulbs and the size of the cerebral hemispheres, but more primitive than that of any modern lemurs in the size of the brain as a whole and the relative size and exposure dorsally of the hind-brain. It may be noted that if the Plesiadapidæ are primitive lemuriforms, they carry back the evolutionary development of the lemuroid brain to a much earlier stage, for the remains of the skull of *Stehlinella* (see Fig. 12) indicate that in this form the brain was quite small, with large olfactory bulbs and a poorly developed neopallium, resembling thus the brain of generalized and lowly insectivores.

It is clear, at least, that the brain of the recent Lemuroidea in its most primitive expression is of a simple type, though showing many advances over brains of lower mammals of an equivalent body-weight not only in its size but in the expansion of the neopallium as evidenced by the deep Sylvian fissure, the well-developed calcarine complex of sulci, and the relative extent of the " association areas."

In the larger lemurs the neopallium is more richly convoluted, and these forms indicate the prevailing tendencies in the production of a fissural pattern characteristic of the Lemuroidea. It has been demonstrated very conclusively that the development of cortical sulci is in great part related to the expansion of the different cortical areas. For some sulci (limiting sulci) are produced along the line of junction between two adjacent areas, while others (axial sulci) arise as a folding down the middle of one area, allowing thus

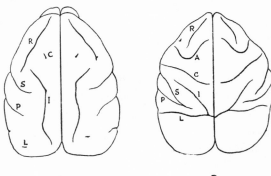

A   B

FIG. 49.—DORSAL VIEW OF THE BRAIN OF A, *Lemur nigrifrons* × 1 AND B, *Cercopithecus mona* × ⅔, TO SHOW THE DIFFERENT DISPOSITIONS OF THE CORTICAL SULCI.

A, Arcuate sulcus ; C, Central sulcus ; I, Intraparietal sulcus ; L, Lunate sulcus ; P, Parallel sulcus ; R, Sulcus rectus ; S, Sylvian fissure.

of an expansion of that particular cortical area. In the typical lemuroid brain there is a striking absence of limiting sulci in the frontal and parietal regions. On the contrary, the sulci which develop here tend to run longitudinally, cutting right across the series of cortical areas which are represented as strips of cortex disposed more or less transversely. Thus the sulcus rectus runs across the middle of the prefrontal (8), frontal (6) and motor (4) areas, while the sulcus intraparietalis likewise cuts across the several areas of the parietal lobe and may run longitudinally into the area striata.* There

* Reference should be made to the chart of the cortical areas in the brain of *Lemur*, published by Brodmann.[1]

may or may not be a small sulcus post-lateralis (sulcus lunatus) lying in the anterior boundary of the area striata on the lateral surface of the brain. The sulcus centralis (which is so conspicuous an element in the brains of higher Primates) is represented by a very small dimple near the upper end of the boundary between the motor and general sensory areas. In the extinct *Lemur jullyi*, the central sulcus was fairly long, but disposed sagittally and not transversely as in the Anthropoidea.[8]

Although this type of fissural pattern predominates among the Lemuroidea, there is a conspicuous exception in the brain of *Perodicticus*, for in this genus a distinct and long sulcus centralis is developed, strictly comparable with the sulcus centralis of monkeys (as shown by Elliot Smith, Brodmann and Vogt). An approximately similar condition is sometimes found in *Propithecus*, though in this case the presumed central sulcus does not consistently form the caudal boundary of the motor cortex as in *Perodicticus* and the Anthropoidea. It is generally held that a true

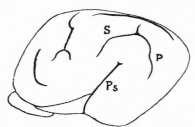

Fig. 50.—Cerebrum of the Aye-aye (*Chiromys*) × 1.
(After Elliot Smith.)

P, Post-Sylvian (parallel) sulcus; Ps, pseudo-Sylvian sulcus; S, supra-Sylvian sulcus.

sulcus centralis is only to be found among the Primates, and hence its appearance in the Potto is of particular significance in considering the place of the Lemuroidea in the order of the Primates. However, there is some evidence (of a histological nature) that the so-called sulcus orbitalis of the primitive insectivore *Gymnura* is, at least in part, the equivalent of the central sulcus, and thus this sulcus may really be more primitive than has hitherto been supposed.

It has been noted above that in the brain of the Aye-aye the supra-Sylvian sulcus does not extend down to meet the rhinal sulcus and does not merge with the pseudo-Sylvian sulcus. This feature, in which it is more primitive than the other Lemuroidea, is apparently associated with a poor development of the parietal and temporal regions of the cortex. More interesting in this brain, however, is the continuity of the supra-Sylvian sulcus with the post-Sylvian sulcus to form a supra-Sylvian arc which curves over the upper end of the pseudo-Sylvian sulcus.* This pattern reproduces the condition commonly found in lower mammals such as the Carnivora. The striking contrast between the fissural pattern in *Chiromys* and other lemuroids can hardly be explained except by supposing that they diverged from each other at a remarkably early stage in evolution, taking origin (it is to be inferred) from a common ancestral type which still preserved potentialities for cortical development either along the typical Primate line or along the line followed by many members of other mammalian Orders.† In the brain of such a type the true Sylvian fissure had presumably not yet developed, and if the supra-Sylvian and pseudo-Sylvian sulci were present at all at this stage they must have been separate. To this extent at least, therefore, the brain of the protolemuroid stock was certainly more primitive than that of *Microcebus*.

Tarsioidea

In spite of (or perhaps because of) the rarity of *Tarsius*, the brain of this small Primate is better known in its minute structure than the brains of most mammals. It is very remarkable on account of its unique combination of advanced with primitive characters.

---

* A similar fissural pattern was apparently present in the fossil lemur *Palæopropithecus*, judging by its endocranial cast (Elliot Smith).

† The sulci in the frontal lobe of the Aye-aye are extremely variable, and their interpretation is a matter of great difficulty. In some cases (*e.g.*, in that shown in Fig. 50) two small elements may link up to form a continuous sulcus which Elliot Smith believes is possibly the equivalent of the sulcus centralis of higher Primates.[6]

Viewed from above, the cerebral hemispheres are very broad and rounded, while from the lateral aspect they show a marked posterior prolongation so that the occipital lobes overlap to a considerable extent the cerebellum and medulla, recalling the characteristic condition of the Anthropoidea. The olfactory bulbs are more reduced than in the Lemuroidea, and, coincidently, the olfactory tract, olfactory tubercle and piriform lobe are all small. It should be noted, however, that the latter appears on the lateral surface of the hemisphere

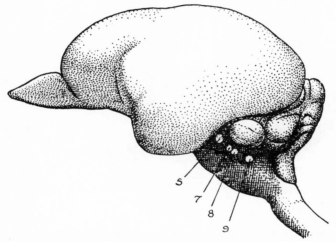

FIG. 51.—LATERAL VIEW OF THE BRAIN OF *Tarsius*.
(Woollard, *P.Z.S.*, 1925.)

forming the tip of the temporal pole as it does in the lemuroid brain. The cerebral hemispheres show no fissuration on their exposed surface except for a short oblique sulcus which may be present in the Sylvian fossa and which possibly represents a rudimentary supra-Sylvian sulcus. The Sylvian fossa (which may be regarded as a very broad pseudo-Sylvian sulcus) is excavated by the large orbits over the roof of which the brain is moulded. Thus *Tarsius* lacks a well-developed supra-Sylvian sulcus, and, of course, there is no true Sylvian fissure such as is found in the brains of other Primates. On the medial surface of the occipital lobe of the hemisphere is a deep triradiate calcarine sulcus, similar to the corresponding sulcus in *Microcebus* and lying in a depression which fits over the large colliculi of the mid-brain. The corpus callosum is short and very primitive in appearance, resembling indeed the same structure in the hedgehog (*Erinaceus*), and thus quite different from the elongated commissure which is so characteristic of Primates generally. On the other hand, the ventral commissure is relatively large, and here again *Tarsius* stands out in contrast to all other Primates of which the brain

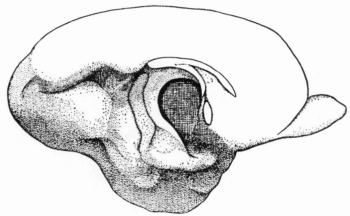

FIG. 52.—MEDIAL VIEW OF THE BRAIN OF *Tarsius* (× 4).

is known. The fornix commissure is small and the hippocampal formation rather inconspicuous. The dentate gyrus is a slender and even strip of cortex reaching from the corpus callosum down to the base of the brain, while the hippocampus is only exposed below.

The structure of the cerebral cortex of *Tarsius* has been studied by Woollard, and perhaps the most striking result of this study is the demonstration of the extent and complexity of the visual cortex (area striata). The expansion of this area is no doubt associated with the enormous size of the eyes, and has also been the cause of the great backward extension of the occipital lobe over the hind-brain. But, apart from its surface extent, the visual cortex appears to be unique amongst Primates in its degree of histological differentiation. It has been reported above that characteristically in the Primates the area striata shows (among other features) a double layer of granule cells with an intervening layer (or stria) of fibres. In *Tarsius* there are three well-defined layers of granules with two fibre layers intervening.[2] In other words, the white stria has itself been split into two by a layer of granule cells. This degree of differentiation and complexity in the area striata is approached but not quite paralleled in certain of the Platyrrhine monkeys (*e.g.*, *Cebus*), and surpasses considerably the condition found in Old World Primates, including even Man. In the development of other cortical areas *Tarsius* approximates to *Microcebus* except that the temporal areas are definitely smaller, while the frontal cortex (area 6) also seems to be rather less in extent.

As regards the ventricular system, it may be noted that the large occipital lobe is hollowed out by a posterior horn of the lateral ventricle, an extension which is present in the Anthropoidea but absent in the Lemuroidea.

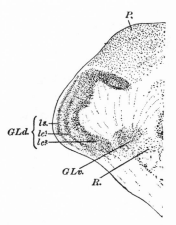

Fig. 53.—Transverse Section through the Lateral Geniculate Body of *Tarsius* to show the Disposition of the Cell Laminæ.

The thalamus of *Tarsius* is similar in its general proportions to that of *Microcebus*. The details of the lateral geniculate body are, however, of special interest. This relatively large structure consists of three laminæ only (and not six as in *Microcebus* and in the Anthropoidea generally), it has no large-celled element (which is present in all other Primates), and no small-celled element (which is present in all lemuroids). Thus the geniculate body of *Tarsius* in this combination of characters stands by itself. It is curious that, although in this animal the visual cortex shows such a high degree of differentiation, the lateral geniculate body in the number of its laminæ and in the lack of cellular differentiation is considerably more simple than in most other Primates. In the manner of folding of its laminæ, it conforms with the lemuroid arrangement of which, indeed, it seems to represent an extreme degree. For, if attention is given to the cross-section of the nucleus shown in Fig. 53, it will be evident that the laminæ have curled inwards so as to form a deep medial and anterior concavity from which the optic radiations emerge, while the fibres of the optic tract largely enter its exposed convex surface. In other words, the lateral geniculate body is markedly inverted.* The importance of this observation lies in the fact that it is hardly possible to postulate the derivation of the completely everted type of geniculate body found in the monkeys from the completely inverted tarsioid type (*vide infra*, p. 168).

The mid-brain of *Tarsius* shows a primitive feature in the relatively enormous size of the colliculi. The cerebellum is very simple, with lateral lobes which are even smaller and less fissured than in *Microcebus*. The simplicity of this part of the cerebellum, indeed, is rather out of keeping with the relative extent of the neopallium and is surprising in view of the animal's agility among the branches. The pons is correspondingly feeble in its development.

We may now attempt to evaluate the significance of the advanced and primitive features of the brain of *Tarsius*. In regard to the former it may be noted that, judging from the skull of *Necrolemur*, the brain of this fossil tarsioid was rather less pithecoid in appearance, for it is clear that not only

* When I first made a study of the thalamus of *Tarsius*[3] I did not appreciate the significance of this feature, and also, not realizing the extent to which the geniculate body had undergone rotation in this animal, I misinterpreted the appearances which it showed in microscopical sections. I reported, therefore, that the lateral geniculate body of *Tarsius* resembles that of the Anthropoidea. Later[4] I was able to establish that this resemblance is really quite spurious.

were the cerebral hemispheres less expanded, but they did not overlap the cerebellum to quite the same extent. Since we know from fragmentary skeletal remains that *Necrolemur* had already acquired the highly peculiar specializations of the hind-limbs, it is not improbable that at least some of the advanced characters in the brain of the living tarsier have been developed independently of the Anthropoidea. The large size of the occipital lobes and the structural differentiation of the visual cortex in *Tarsius* may be entirely associated with the elaboration of the eyes for a nocturnal mode of life and do not necessarily, therefore, indicate a close affinity with the monkeys in which these features of the brain are well established. As shown above, moreover, the structure of the visual cortex in *Tarsius* exhibits unique characters. Again, the reduction of the olfactory bulbs (and other olfactory centres) may be indirectly secondary to the enormous expansion of the orbits which has led to a marked compression of the nasal cavities. On the other hand, the preservation of certain unusually primitive traits in the tarsioid brain indicates that this sub-order has been derived phylogenetically from quite a generalized mammalian type. In the conformation and relative size of the commissures, in the absence of a true Sylvian fissure, and in the simplicity of the cerebellum, the brain is much more primitive than that of any recent lemuroid. Even the Eocene *Adapis* evidently possessed a deep Sylvian fissure of the Primate type, with large expanded temporal lobes which in *Tarsius* are but slightly developed.

If we attempt to construct a mental picture of a mammalian ancestral type common to the recent Lemuroidea and the Tarsioidea, we must infer that in the brain of such a form the neopallial cortex would be quite smooth except for a pseudo-Sylvian fissure and possibly a calcarine sulcus, the corpus callosum of a very primitive eutherian type, the ventral commissure rather large, the occipital pole of the hemisphere not produced backwards to overlap the hind-brain, and the cerebellum at least as simple as it is in *Tarsius*. Such a brain would evidently have exhibited no features by which it might be labelled as "lemuroid" rather than "primitive tarsioid." The evidence provided by the brain of *Chiromys*, indeed, suggests that the Lemuroidea commenced a divergent evolutionary trend at a very early stage when the common progenitor of the Primates was little more than a generalized eutherian mammal with a small brain of insectivorous type, and if the Plesiadapidæ are accepted as early precursors of the Chiromyidæ, this conclusion becomes unavoidable.

It has been pointed out that the brains of certain fossil tarsioids were evidently more primitive in their general contour than those of *Tarsius*.   On the other hand, it was long ago demonstrated by Cope that the brain of the Eocene species *Tetonius homunculus* ($=$ *Anaptomorphus homunculus* of that author) was quite as large as that of the modern tarsier.*   This observation is of interest from two points of view.   It shows that the Primates at a very early geological time embarked upon a line of evolution characterized by a progressive enlargement of the brain, for in contemporaneous Eocene mammals of carnivorous and ungulate stocks the brain was still astonishingly small as compared with their modern representatives. It indicates also that, so far as cerebral development is concerned, *Tarsius* has remained stationary since almost the beginning of Tertiary times.   Lastly it may be emphasized that, in spite of pithecoid features such as the extensive occipital lobes and the breadth of the cerebral hemispheres, the brain of *Tarsius* as a whole is not appreciably larger in proportion to body-weight than that of *Nycticebus*.   Thus, as Elliot Smith[9] has emphasized, "the Tarsioidea are no better off than the lemurs so far as the quantity of the brain is concerned," or, in other words, the cerebral status of these two Primate sub-orders is of an equivalent order.

---

* On the other hand, the brain of the Eocene Lemuroidea was evidently much smaller relative to body-weight than in their modern representatives (with the possible exception of *Aphanolemur*, a fossil lemuriform of North America whose status is somewhat uncertain).

Anthropoidea

The brain in the smaller monkeys shows a strong contrast with that of the lower Primates in its size. Elliot Smith has shown that, comparing animals of approximately the same body-weight, the brain of the Platyrrhine monkey is three to four times as large as that of a lemuroid. By far the most primitive type of brain in the Anthropoidea is represented in the Hapalidæ, and perhaps this provides the strongest argument in favour of the conception that these small monkeys are truly primitive forms, for there is no evidence at all that the simplicity of their brains is the result of secondary retrogression.

In *Hapale* the cerebral hemispheres are voluminous with conspicuously large frontal lobes, and a backward extension of the occipital lobes which overlap to a very considerable extent the cerebellum and medulla. This latter feature reaches its maximum expression among the monkeys in *Chrysothrix*. The olfactory bulbs are greatly reduced (more so than in *Tarsius*), and there is a corresponding diminution in the olfactory tubercle and in the piriform lobe. The piriform cortex is almost entirely displaced by the expanding neopallium on to the basal and medial surface of the brain, and is limited by a shallow rhinal sulcus which can be traced quite readily throughout its usual course. The corpus callosum (in correlation with the neopallial expansion) is relatively large and elongated, with a well-marked splenium and genu, while the ventral commissure and the fornix commissure are both very small. The dentate gyrus is a narrow crinkled band of cortex reaching from the splenium of the corpus callosum down to the base of the brain, and the hippocampus is only exposed at its lower extremity (as in *Tarsius*, but to a slightly less extent).

The neopallial cortex is smooth except for a Sylvian fissure, a faintly indicated parallel sulcus, and the calcarine sulcus.

The Sylvian fissure is of the Primate type—that is to say, it is formed by the supra-Sylvian sulcus which reaches down to the rhinal sulcus and blends there with the pseudo-Sylvian sulcus. The parallel sulcus (=post-Sylvian or superior temporal sulcus) is quite short and very shallow, so that it usually forms

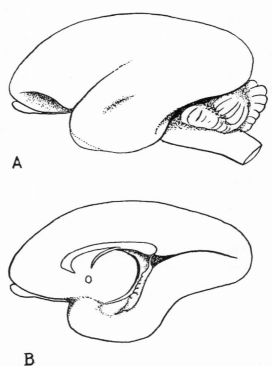

A

B

FIG. 54.—LATERAL, A, AND MEDIAL, B, VIEWS OF THE BRAIN OF THE MARMOSET (*Hapale jacchus*) × 2.

little more than a dimple on the surface of the temporal lobe. The calcarine fissure is comprised almost entirely by the retrocalcarine element which is very deep and thus provides for a marked expansion of the visual cortex in the longitudinal axis of which it is situated. The most anterior extremity of this fissure probably represents a short precalcarine sulcus, while the sulcus paracalcarinus (found in *Tarsius* and lemurs) is absent. It is important to note that the relative smoothness of the neopallium in the marmoset—as compared with the brains of other Primates—can hardly be attributed altogether to the small size of the animal, for in small lemuroids such as *Galago* the cerebral cortex is relatively quite well convoluted.

An analysis of the cortical structure in the marmoset shows that, compared with lower Primates, there is a considerable expansion of the "association areas," especially in the frontal and parietal regions. In the latter the visual and general sensory areas are now widely separated, while in the former the prefrontal cortex is more obtrusive. The visual cortex is well differentiated in all the Anthropoidea, showing in general a double layer of granule cells with an intervening fibre layer. In some of the Platyrrhines, however (*e.g.*, Cebidæ), the differentiation of the area striata has proceeded further than in the Catarrhine monkeys and even the anthropoid apes and Man, for the fibre layer tends to be split into two by the appearance in the middle

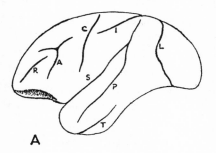

A

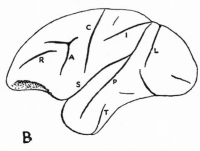

B

FIG. 55.—THE CEREBRUM OF A PLATYRRHINE MONKEY, *Cebus,* A, AND A CATARRHINE MONKEY, *Macacus,* B, TO SHOW THE ESSENTIAL IDENTITY OF THE ARRANGEMENT OF THE CORTICAL SULCI.

A, Arcuate sulcus; C, Central sulcus; I, Intraparietal sulcus; L, Lunate sulcus; P, Parallel sulcus; R, Sulcus rectus; S, Sylvian fissure; T, Inferior temporal sulcus.

of it of a third layer of cells. As already pointed out, the Cebidæ make an approach to the specialized visual cortex of *Tarsius* in this regard.

In most of the Old and New World monkeys the cerebral

cortex becomes richly convoluted (Fig. 55). Especially conspicuous are the sulcus centralis, which forms an accurate boundary-line between the motor cortex and the general sensory cortex, and the sulcus lunatus, which limits anteriorly the area striata on the lateral surface of the brain. In addition, the intraparietal sulcus separates the area parietalis (7) from the area preparietalis (5), while the arcuate sulcus separates the area frontalis agranularis (6) from the area frontalis granularis (8). Thus, on the dorso-lateral surface of the hemisphere the sulci in general are disposed transversely as limiting sulci which develop at the junction of the transversely arranged cortical areas (Fig. 49). Herein is rather a striking contrast with the prevailing condition in the Lemuroidea, for in this group (as shown above) the sulci which appear in this region tend to be disposed longitudinally, cutting across these areas instead of separating them (although exceptions are to be noted in *Perodicticus* and occasionally in *Propithecus*). The fissural pattern of the larger Platyrrhines is remarkably similar to that of the Catarrhine monkeys, a fact to which attention has been called by several observers (Fig. 55, A and B), and yet (judging from the constancy of certain differences in the skull and dentition, etc.) it is to be presumed that these two groups separated in their evolutionary development quite early, even if they did not arise independently from two separate tarsioid stocks. It is at least tolerably certain that the brain of the common ancestor of the Old and New World monkeys could not have been less primitive than that of the Hapalidæ, and in this case the identity of the fissural patterns in the larger Platyrrhine and Catarrhine brains must be the result of evolutionary parallelism. This pattern can also be traced in its fundamental elements in the brains of the anthropoid apes and Man, but it becomes increasingly obscured and complicated in them by the addition of secondary sulci which serve greatly to increase the surface area of the cortex. This expansion involves almost entirely the " association areas," for within the limits of the Anthropoidea there is but little relative variation of the histological extent of the sensory and motor areas. With the progressive elaboration of the parietal "association areas" the occipital pole of the hemisphere becomes more pronounced and the visual area pushed further back, and ultimately—in the human brain—this area becomes almost entirely accommodated on the medial surface of the brain.

The thalamus of the Anthropoidea is considerably more

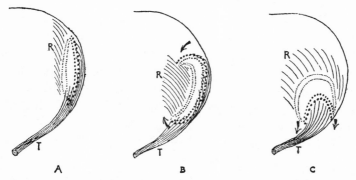

A   B   C

Fig. 56.—Diagram of Transverse Sections through the Lateral Geniculate Body, showing the Changes which Occur in the Cell Laminæ in the Primates.

In A is shown a generalized condition such as is seen in primitive lemuroids in which the laminæ are vertically disposed, the optic tract fibres, T, entering from below and laterally, and the optic radiations, R, leaving the medial aspect. In Lemuroidea generally and in *Tarsius*, the laminæ become inverted, B, while in all the Anthropoidea they become everted, C.

advanced in its development than in lower Primates. The expansion of the lateral nucleus has broadened out the dorsal surface and has also pushed back the caudal part of this nucleus (the pulvinar) to form a prominent rounded projection. It is partly for this reason that the pulvinar has been regarded by some comparative neurologists as a characteristic Primate feature. Actually, however, it may be quite a large element in the thalamus of lower mammals, but it is not in them extruded into such a conspicuous position by the growth of the lateral nucleus. The dorso-medial nucleus in the pithecoid thalamus is large and intrinsically differentiated in correlation with the expansion and differentiation of the prefrontal cortex.

The lateral geniculate body of all Anthropoidea is characterized by the fact that its constituent laminæ have become everted to a varying degree so that they form a ventrally directed concavity or " hilum " at which the fibres of the optic tract enter, and a dorso-medial convexity from which the optic radiations emerge (Fig. 56). The cell laminæ are usually six in number, and the two superficial laminæ are composed of large cells. There is no small-celled element as in the Lemuroidea. In some monkeys, *e.g.*, *Macacus*, the lamination of the lateral geniculate body is even more complex than in Man, for in this species two additional laminæ may be separated off in the area which is concerned with receiving fibres from the macula of the retina, and thus in this region there are no less than eight layers of cells. The eversion of the lateral geniculate body is complete in all the monkeys and also in the gibbon. In the other anthropoid apes and in Man it is not complete, for the lateral extremities of the laminæ project to form a spur or tail (*cauda* of lateral geniculate body, Fig. 47).

The mid-brain in *Hapale* shows relatively large colliculi. The cerebellum is represented here in its simplest form among the Platyrrhines, but it is considerably more elaborate than that of *Tarsius*, the lateral lobes especially being larger and more complexly fissured. In the Catarrhines and higher Primates a " paramedian lobule " is differentiated from the lateral lobe, but this is said to be absent in the New World monkeys as it certainly is in *Tarsius* and *Microcebus*. According to Bolk this lobule is present in most of the Lemuroidea.

In considering the evolutionary origin of the pithecoid brain, we may employ the hapalid brain for our discussions, for, as we have seen, this is the simplest and most primitive expression of the brain in the living Anthropoidea. Clearly it shows many great advances over the brains of lower Primates of an equivalent body-weight in its absolute size, in the expansion of the cerebral cortex, and in the complexity of the cerebellum. It would appear, however, that the main factor in the production of a simian from a prosimian type of brain is related to the expansion of the so-called " association areas " of the cortex, which become increasingly differentiated histologically. These areas comprise a neural mechanism which allows the animal to react to environment stimuli with greater precision and a wider variety of possible adjustments, and makes possible an enhanced behavioural plasticity and increasing adaptability to changes in the environment.

The brain of *Tarsius* in several features seems to foreshadow the pithecoid brain—*e.g.*, in the relative size and conformation of the cerebral hemispheres (and especially the marked backward extension of the occipital lobe), in the differentiation of the visual cortex, in the reduction of the olfactory centres, and in the presence of a posterior horn of the lateral ventricle. But there are also some fundamental differences apart from the fact that in its cerebral commissures, cerebellum, etc., the brain as a whole is very much more primitive than any simian brain. Attention has already been drawn to the differentiation and the manner of folding of the lateral geniculate body. In all the Anthropoidea this nucleus becomes everted, while in *Tarsius* it is completely inverted. A reference to the diagrams in Fig. 56 will indicate that in these processes of inversion and eversion the geniculate bodies of *Tarsius* and the Anthropoidea have developed in diametrically opposite directions. Hence it would seem only reasonable to infer that the geniculate bodies in these two groups have been derived from a simple type (such as is seen in the brain of the tree-shrew) in which the laminæ remain still unfolded in either direction. The absence of a large-celled element in the geniculate body of *Tarsius* (which is present in all other Primates) provides further evidence that a common ancestral type must have possessed a structure of quite a simple and unelaborated form. If this is so, it seems certain that the brain of a primitive Primate which might have given rise to both the tarsioid and the anthropoid types could not have exhibited the expansion and differentiation of the visual cortex which is to be seen in the modern *Tarsius*, and thus probably did not show pithecoid features such as the enlargement and extension backwards of the occipital lobe of the hemisphere or the development of a posterior horn of the lateral ventricle. Hence it is apparent that these advanced characters of the brain of *Tarsius* probably developed independently of the evolution of the higher Primates, as the result of parallelism.

But although the evidence of brain structure indicates a very early separation of the tarsioid and anthropoid stocks, the fact that such conspicuous pithecoid features have become manifested in the brain of *Tarsius* (the structure of the visual cortex closely resembling that of certain Platyrrhine monkeys even though the lateral geniculate body has undergone differentiation in a very different direction) provides convincing evidence that *Tarsius* and the Anthropoidea at least had their origin in a common ancestral stock with potentialities for evolutionary development along these similar lines.

It has been demonstrated that the lateral geniculate body displays a somewhat more primitive feature in Man and the large anthropoid apes than it does in the gibbon or in the monkeys, for in the former its laminæ are not completely everted. This interesting fact may indicate that the whole pithecoid stock represented by the living Platyrrhine and Catarrhine monkeys is to that extent more specialized, and thus could hardly have formed a stage in the evolution of Man and the larger apes. This interpretation is open to question, for the size of the brain as a whole may be one of the factors which, in the Anthropoidea, determines the degree of eversion of the geniculate body.*

* But in any case, of course, a complete eversion of this nucleus need not be taken as an essential diagnostic character in the definition of the terms *Catarrhine* or *Platyrrhine*, even though it is universally found in living members of these pithecoid groups.

The remarkable similarity of the human brain to that of the anthropoid apes is well known. The differences which it shows in comparison with the gorilla brain are entirely quantitative, and the multitudinous qualitative resemblances extend to the smallest details. And yet, if the evidence of foot structure is accepted (*vide supra*, p. 131), it would seem that the human stem must have diverged from the stem which culminated in the living anthropoid apes at a stage when the body-weight had hardly exceeded that of the little gibbon. Thus there is evidence that many of the human features and proportions of the gorilla brain have developed independently of the evolution of Man. Nevertheless, there can be no doubt on the basis of comparative anatomy—and the endocranial casts of fossil Hominidæ such as *Pithecanthropus* and *Sinanthropus* furnish remarkable corroborative evidence—that in its evolutionary development the human brain passed through a phase in which it was practically identical with that of the modern gorilla. In other words, if we limit our observations to the brain, we can only infer that among the progenitors of the human race was a form which would be quite legitimately classed as an " anthropoid ape."

References

1. BRODMANN, K.: Die Cytoarchitektonische Cortexgliederung der Halbaffen. Journ. für Psych. u. Neur., 1907.
2. CLARK, W. E. LE GROS: The Visual Cortex of Primates. Journ. of Anatomy, vol. lix., 1925.
3. CLARK, W. E. LE GROS: The Thalamus of Tarsius. Journ. of Anatomy, vol. lxiv., 1930.
4. CLARK, W. E. LE GROS: The Brain of Microcebus. Proc. Zool. Soc., 1931.
5. CLARK, W. E. LE GROS: The Lateral Geniculate Body. Brit. Journ. of Ophthalm., vol. xvi., 1932.
6. ELLIOT SMITH, G.: Catalogue of the Physiological Series of Comparative Anatomy in the Royal College of Surgeons, vol. ii., 1902.
7. ELLIOT SMITH, G.: On the Morphology of the Brain in the Mammalia. Trans. Linn. Soc., vol. viii., 1903.

8. ELLIOT SMITH, G. : On the Form of the Brain in the Extinct Lemurs of Madagascar. Trans. Zool. Soc., vol. xviii., 1908.

9. ELLIOT SMITH, G. : Discussion on Tarsius. Proc. Zool. Soc., 1919.

10. ELLIOT SMITH, G. : Essays on the Evolution of Man. London, 1927.

11. GREGORY, W. K. : On the Structure and Relations of Notharctus. Mem. Amer. Mus. Nat. Hist., vol. iii., 1920.

12. NEUMAYER, L. : Über das Gehirn von Adapis parisiensis. Neues Jahrb. f. Mineral, vol. ii., 1906.

13. WOOLLARD, H. H. : The Anatomy of Tarsius spectrum. Proc. Zool. Soc., 1925.

14. WOOLLARD, H. H. : The Cortical Lamination of Tarsius. Journ. of Anatomy, vol. lx., 1925.

CHAPTER VII

THE EVIDENCE OF THE SPECIAL SENSES

PERHAPS the most conspicuous feature in which the Primates as a group stand in contrast with other mammalian Orders is manifested in the elaboration and refinement of those special sense organs and their neural connections which form the basis of the faculty of discrimination. By the enhancement of these mechanisms the animal is enabled to define with much more precision the varying influences of its environment, and to react with a much greater accuracy thereto. Moreover, it seems not improbable—as Elliot Smith has pointed out— that the progressive development of these senses provides a stimulus which largely conditions the evolutionary expansion of the brain as a whole.[6]

The visual sense is by far the most important of the discriminative senses, for it provides a means whereby objects may be recognized near at hand or in the far distance in regard to their precise position in space, their colour, texture and form, and enables these properties to be defined with an accuracy which is hardly approached by the use of other sensory mechanisms. On the other hand, the sense of smell permits of very little discrimination except in reference to the quality and intensity of the odour of an object, and provides no information as to its other qualities or even (to more than a very slight degree) to its position in relation to the observer. With the increasing perfection and domination of the more discriminative sense organs, therefore, the sense of smell becomes of less importance, and, coincidently with this, the olfactory organ undergoes a progressive atrophy. This retrogression, which is probably the indirect result of an arboreal habitat, has been noted in the sections dealing with the skull and brain, where it is related to the recession of the snout region and the shrinkage of the olfactory bulbs, etc.

The Olfactory Organ

In most lower mammals the external nares or nostrils are surrounded by an area of naked and moist glandular skin which constitutes what is properly called the rhinarium. Possibly this moist surface aids the purely olfactory sense by enabling the animal to detect the direction of air-currents which carry an odour from a distance. Primitively the rhinarium is prolonged down from the nostrils over the median part of the upper lip, to become continuous with a stout frenum which binds the lip firmly to the underlying gum. This condition is shown in Fig. 57, A. It may be recalled that during embryological development various processes grow forwards

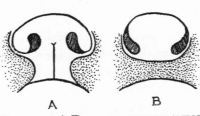

A   B

FIG. 57.—A DIAGRAM ILLUSTRATING THE STRUCTURE OF THE EXTERNAL NOSE AND UPPER LIP IN A, THE LEMUROIDEA AND B, *Tarsius* AND THE ANTHROPOIDEA.

(After J. D. Boyd, *Journ. Anat.*, 1932.)

The maxillary processes are represented by the stippled area.

and downwards on either side from the region of the base of the skull and, by their ultimate fusion, construct the whole of the face. Thus the median nasal process which occupies the mid-line forms the septum of the nose and the labial portion of the rhinarium, while on either side maxillary processes (represented in Fig. 57 by the stippled area) extend towards the mid-line to meet the margins of the median nasal process and to form the cheek region and the hairy lateral part of the upper lip. In the Lemuroidea this primitive mammalian type of rhinarium persists. The maxillary processes do not meet in the mid-line, and thus the median nasal process is left exposed in the middle of the upper lip to form the labial part of the rhinarium.   The latter is furrowed by a median groove which represents the union of two parts of the originally bifid median nasal process.   Occasionally, as in *Galago*, the maxillary processes may extend further medially so as to come into a direct contact with each other, so that, in this case, the labial portion of the rhinarium is sunk into a deep groove and the whole of the superficial parts of the upper lip is covered with hairy skin.[1]

In *Tarsius* and all the Anthropoidea the structure of the

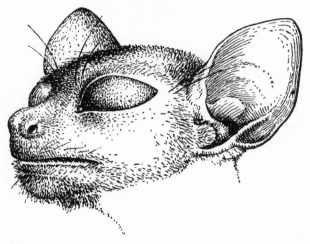

FIG. 58.—THE FACE OF *Tarsius* TO ILLUSTRATE THE PITHECOID APPEARANCE OF THE NOSTRILS AND UPPER LIP.   (Woollard, *P.Z.S.*, 1925.)

external nose is rather different (Figs. 57, B, and 58).   The labial part of the rhinarium is no longer present, and it appears that the maxillary processes have completely fused in front of the median nasal process so that the latter is buried from the surface (Boyd).*   Thus the upper lip shows a continuous smooth surface uninterrupted by a median groove.   With

* The fact that the superficial part of the middle of the upper lip (the philtrum) is really derived from the maxillary processes, and not from the median nasal process, was demonstrated by J. E. Frazer in the development of the human embryo.

these changes are associated (1) a reduction of the elements of the upper lip formed from the median nasal process, including the attenuation of the frenum, and (2) the muscularization of the whole of the upper lip across the mid-line, the muscle elements having been carried to this position by the extension of the maxillary processes. These two features lead to an increased freedom and mobility of the upper lip which characterize all the Primates with the exception of the Lemuroidea. The type of rhinarium shown in Fig. 57, B is also found in some lower mammals (*e.g.*, the horse or the pig), but in *Tarsius* or the Anthropoidea the area of naked moist skin round the nostrils

a   b

FIG. 59.—THE FACE OF *a*, A CATARRHINE MONKEY, *Cercopithecus*, AND *b*, A PLATYRRHINE MONKEY, *Cebus*, ILLUSTRATING THE DIFFERENT DISPOSITION OF THE NOSTRILS IN THESE TWO GROUPS.

has entirely disappeared, so that the rhinarium can hardly be said to be present at all. Boyd suggests that in these Primates " the process of maxillary overlapping has become so great that the nasal portion of the rhinarium has become covered over in a manner corresponding to the covering over of the labial portion." The disappearance of the rhinarium proper in higher Primates is doubtless an outward manifestation of the retrogression of the olfactory apparatus as a whole in these forms. Herein a strong contrast is shown between the Lemuroidea on the one hand, and *Tarsius* and the Anthropoidea on the other, and the relatively primitive status of the former is emphasized.

In the monkeys the disposition of the nostrils forms the basis of the distinction between " Platyrrhine " and " Catarrhine." In the Platyrrhine monkeys the nostrils are relatively wide apart and separated by a broad septum, while in the Catarrhines the septum is relatively narrow (Fig. 59). This distinction between New and Old World monkeys is not absolutely sharp, for in some of the former (*e.g.*, *Ateles*) the nostrils are fairly close together, while in the latter (*e.g.*, *Colobus*) they may be almost as wide apart as in a typical Platyrrhine. Moreover, the superficial differences in the position of the nostrils are not reflected at all in the nasal skeleton.

The changes in the external nose of the Primates are paral-

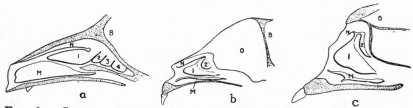

FIG. 60.—LATERAL VIEW OF THE NASAL CAVITY IN *a*, *Lemur mongoz*; *b*, *Tarsius* (after Woollard); AND *c*, *Hapale* (after Beattie).

B, Brain cavity ; M, maxillo-turbinal ; N, naso-turbinal ; 1, 2, 3, 4, endo-turbinals.

leled in the structure of the nasal cavities. In the Lemuroidea these cavities, in association with the prominence of the snout, are relatively large. They are also greatly complicated by the labyrinthine system of turbinate bones which form a series of scroll-like structures projecting into the cavity from its lateral wall. These turbinate bones or olfactory scrolls are folded upon themselves and, being covered on their surfaces by mucous membrane, clearly provide for a considerable extension of the lining epithelium of the olfactory chamber. In general there is a close correlation between the degree of complexity of the turbinate system and the acuity of smell, and there is little doubt that a well-developed system is a primitive mammalian character. In lowly mammals the olfactory scrolls are relatively numerous, and among the edentates there may be as many as nine (*e.g.*, in *Orycteropus*).

In Fig. 60, *a* is indicated the lateral wall of the right nasal cavity in *Lemur*, and the condition shown here represents the arrangement found in almost all members of the Lemuroidea. Most of the olfactory scrolls are developed as processes of the ethmoid bone and are called the ethmo-turbinals. The most anterior—designated the naso-turbinal—extends along the under surface of the nasal bone. Behind this are four principal or endo-turbinal scrolls (of which the first is relatively large) lying one behind the other and reaching back to the sphenoid. On a lower level is the maxillo-turbinal, which is a distinct skeletal element and ossifies from a separate centre. This bone corresponds to the inferior turbinate bone of human anatomy. In Fig. 61 is shown a schematic diagram representing a section across the nasal cavities immediately below and parallel with the cribriform plate of the ethmoid. This diagram shows clearly the arrangement of the naso-turbinal and endo-turbinals in the lemuroid skull. Laterally (and overhung by the endo-turbinals) are seen two small accessory scrolls or ecto-turbinals (E^1 and E^2). Thus in the Lemuroidea generally there are present—besides the naso-turbinal and maxillo-turbinal—four endo-turbinals and two ecto-turbinals. In *Chiromys* a more primitive condition is found, for in this genus there are five endo-turbinals and at least three ecto-turbinals (Kollmann and Papin).[7] In the lesser degree of reduction of the turbinate system the Aye-aye stands in rather marked contrast to other lemurs. It may be noted also that the Lemuriformes and Lorisiformes are distinguished from each other in certain features of the nasal fossæ. In the latter the first ethmo-turbinal is very large and actually covers over the maxillo-turbinal, while in the former it is much smaller. In this respect the Lorisiformes are probably the more primitive of these two groups.

In the higher Primates the turbinate system has been reduced to a remarkable degree. In the marmoset (as in other Anthropoidea) the diminution of the snout region, together with the recession of the jaws to a position rather below the front end of the brain case, has led to a constriction of the nasal fossæ (Fig. 60, *c*). The maxillo-turbinal is small and relatively simple in its structure, and only two endo-turbinals remain (the middle and superior turbinate processes of human anatomy). Moreover these scrolls are disposed in a vertical rather than in an antero-posterior series. The naso-turbinal is shrunk to an inconspicuous ridge —the agger nasi—which can be traced into continuity with the middle turbinate process, and only one ecto-turbinal remains as the bulla ethmoidalis lying under cover of the middle turbinate. That the Anthropoidea have been derived from a form in which the turbinate system was certainly more elaborate is indicated by the fact that even in Man occasionally one or two rudimentary scrolls may be found above the superior turbinate process, and in the human embryo five turbinate processes are commonly to be recognized (including the maxillo-turbinal). It is pertinent here to refer to the significant observations of Wislocki on the scent glands of the marmoset.

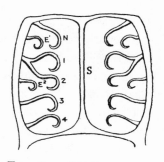

Fig. 61.—Diagram of a Cross-section through the Posterior Part of the Nasal Cavity of a Typical Lemuroid immediately below the Cribriform Plate of the Ethmoid.

S, Nasal septum; N, naso-turbinal; E^1, E^2, ecto-turbinals. 1, 2, 3, 4, endo-turbinals.

According to this authority, this small monkey has a greater complexity of scent glands—which are found mainly in the pubic region and perineum—than any other simian. This indicates that the sense of smell plays an important rôle in the marmosets—a remarkably primitive characteristic in which the Hapalidæ contrast strongly with other Anthropoidea.

The turbinate system in *Tarsius* is very similar to that of the Anthropoidea, for in addition to the small and attenuated maxillo-turbinal there are only two endo-turbinals (Fig. 60, *b*). The naso-turbinal is slender in comparison with the Lemuroidea, but more conspicuous than the agger nasi of monkeys. The reduction of the turbinate system in *Tarsius* is evidently secondary to the restriction of the nasal fossæ, but this restriction has come about in a manner quite different from the analogous changes in the Anthropoidea. For whereas in the latter it is the result of a retraction backwards of the nasal fossæ associated with the recession of the jaws relative to the brain-case, in *Tarsius* it is due to the enormous development of the orbits with the formation of a thin interorbital septum which has seemingly obliterated a large part of the nasal fossæ by lateral compression. Thus, in the latter case, the restricted nasal fossæ are not placed below the front part of the neurocranium as in monkeys, but, on the contrary, are displaced relatively far forward in front of it. This relation is shown clearly in Fig. 60, *b*. Hence it appears that the close resemblances of the turbinate system in *Tarsius* and the Anthropoidea are of a parallel nature—the result of a limitation of space in the nasal cavities which has been conditioned by totally different factors.*

If we consider as a whole the evidence provided by the structure of the peripheral olfactory apparatus, it becomes evident that the Lemuroidea are considerably more primitive than other Primates in this respect, and represent a stage

---

* Kollmann ("Les Fosses Nasales des Tarsiers," *Comptes Rendus Ass. Franc. Sci.*, Congrès du Havre, 1914) cites the observation of E. Fischers that a true interorbital septum is formed during the ontogenetic development of the skull in monkeys, and suggests that the existence of this septum prevents the development of a large number of turbinals as it does in *Tarsius*. But the extreme compression of the nasal cavities by the orbits in *Tarsius* finds no strict parallel among the Anthropoidea. For the purpose of comparison, reference should be made to the figures of the chondrocranium of *Chrysothrix* and *Tarsius* (Abb. 6 and 21) in Henckel's paper, Studien über das Primordialkranium und die Stammesgeschichte der Primaten (*Morph. Jahrb.*, Bd. lix., 1923).

through which the higher Primates must have passed in their evolutionary development. Judging from the condition found in *Chiromys*, they have presumably themselves been derived from a form in which the nasal fossæ contained at least five endo-turbinals and three ecto-turbinals. Thus it is certain that olfaction was strongly developed in the earliest Primates, a conclusion which is well substantiated by the external appearance of the skull in the fossil plesiadapids (see Fig. 12, A).

In *Tarsius* and the Anthropoidea the external nose has undergone a precisely similar development, culminating in the disappearance of a true rhinarium and the acquisition of a freely mobile upper lip. Corresponding resemblances are also to be noted in the olfactory cavities, but in this case they appear to have been produced quite independently. The skull of the fossil tarsioid, *Necrolemur* (Fig. 14), suggests that within the limits of the Tarsioidea there has been a progressive evolutionary change in the conformation of the nasal fossæ from a condition probably similar to that of the recent Lemuroidea. This has involved gradual obliteration of the posterior part of the fossæ by the development of capacious orbits which serve to accommodate the large eyes adapted for a nocturnal mode of life. It cannot be doubted that the inordinate size of the eyes in the modern tarsier represents a high grade of specialization, and thus it is not possible to regard the nasal fossæ of this genus as in any way representing a phase of evolutionary development which could have led on to the structural type of nasal fossæ found in the Anthropoidea. On the contrary, in the latter the nasal fossæ and the features of the turbinate system were apparently also derived from a primitive type such as is found in the Lemuroidea (and probably, indeed, in the early generalized tarsioids in which the orbits were as yet of moderate size), and are directly correlated with the recession of the snout region.

The Eye

In dealing with the structure of the brain, emphasis was laid on the observed fact that in all modern Primates there is a high degree of elaboration of the visual centres, and that in the evolution of this group of mammals the increasing dominance of the visual sense is a characteristic feature. This phenomenon is evidenced also in the structure and differentiation of the sensory epithelium or retina of the eye.

The comparative anatomy of the retina in the Primates has been discussed in a valuable monograph by Woollard,[14] and his observations provide the material for the short discussion which follows.

In the retina of the human eye there are two kinds of percipient elements—the rods and the cones. These two kinds of receptors are commonly found together in the retina of all terrestrial vertebrates,* but in animals which have adopted a nocturnal mode of life as a secondary adaptation the cones disappear. In these cases, therefore, rods only are found. Evidence of an experimental and comparative anatomical nature indicates that the rods function at low degrees of luminosity and probably are mainly concerned with responding to the stimulus produced by movements of objects in the field of vision. They come into action in what has been termed " twilight " or " scotoptic " vision. The cones, on the other hand, are related functionally to higher degrees of visual acuity. They respond to stimuli in a bright light, and form a peripheral mechanism through which it is possible to appreciate colour and the more precise details of form. Whereas several rods are eventually linked up with a single optic nerve fibre, each cone is related to one fibre only. Thus in the case of the cones a more precise conduction of impulses received by the individual elements is allowed. Throughout the greater part of the human retina the rods predominate

* In reptiles, cones are generally predominant in the retina.

over the cones by three or four to one. Towards the periphery the retinal epithelium is almost entirely composed of rods, the cones being here scattered very sparsely. Towards the centre of the retina, however, the cones increase in relative number until at the central point—at the macula lutea or yellow spot—there are practically no rods at all. At the macula there is a small depression, the fovea, where all the layers of the retina disappear except the layer of cones, leaving a small excavation where the latter are brought in direct access to the light. This is the point of greatest visual acuity, for here the light stimuli are enabled to reach the cones directly without the necessity of passing through other retinal layers. At the periphery of the retina the sensory epithelium ceases abruptly, forming a crinkled margin which is called the ora serrata.

In lower Primates there is considerable variation in the structure of the retina dependent partly upon the degree of differentiation of the macular region, but in general there are two main types to be recognized. In one type, the diurnal retina, cones are present and increase in number towards the central point of the retina. In the nocturnal type, on the other hand, rods only are present. As stated above, this latter type is found in animals of nocturnal or crepuscular habits, and it always lacks a fovea. In the Anthropoidea the retina is of the diurnal type with one exception only, that of the Platyrrhine monkey *Nyctipithecus*. If, now, the retina of monkeys and apes is compared with that of the human eye, it is found that in some cases a rather higher degree of differentiation has been attained in the former. Thus Woollard, in assessing the visual acuity by the number of conducting elements in the retina, the degree of intermixture of rods and cones, and the extent and perfection of the fovea, places *Cercocebus* and the chimpanzee before man. In *Cercocebus* there is almost a pure cone retina, occasional rods only being found at the periphery, while the fovea is extremely regular and deep, and the retinal layers other than the cones are here thinned out to a minimum. In the chimpanzee rods are present in the macular region and the fovea is not so fully excavated as in *Cercocebus*. It may be noted that the chimpanzee is the only sub-human Primate in which a sharply defined ora serrata is present (Franz). In the other apes and in all monkeys the anterior limit of the sensory part of the retina is marked by a gradual transition into the non-sensory part.

In the retina of the marmoset the percipient layer in the macular region is composed entirely of cones, but it exhibits a primitive feature in the structure of the fovea. This small pit does not show such a complete excavation as in other Anthropoidea, for the inner nuclear layer of the retina stretches across it so that the cones are not directly exposed to light stimuli. Thus *Hapale* provides an interesting stage in a progressive structural differentiation leading on to the more perfect type of fovea in which the cones become completely uncovered.

The retina of *Nyctipithecus* is unique among all the Anthropoidea in having no macula and no fovea—a fact which was first established by Woollard. The percipient elements of the retina consist entirely of rods, and, except that these become gradually more numerous towards the central part of the retina, the latter shows no differentiation. The retina of this monkey is therefore of the nocturnal type. In *Tarsius* and all Lemuroidea (which have been studied to this end) the retina is also of the nocturnal type—the percipient layer consists entirely of rods and there is no macula.

In the retina of *Tarsius* Woollard[13] has described a local differentiation at the central point which is of peculiar interest. In the region where, in the higher Primates, a macula is found, the retinal epithelium becomes much thickened, the layer of rods is thrown into convoluting folds, and the cells of the ganglion layer are more numerous. This differentiated area (which is circular in outline and measures about 2 mm. in diameter) has been called by Woollard the " primordium maculæ." But it is important to note that it is constructed quite differently from a true macula,* and it can hardly represent a structural stage in the evolutionary development of the anthropoid macula the essential features of which are the complete or almost complete absence of rods and the thinning out of the other layers of the retina. As Woollard has pointed out, the " primordium maculæ " of *Tarsius* bears a close resemblance to the type of retinal differentiation which was described by Kolmer in certain bats (Megacheiroptera).

In the Primates generally, the Lemuroidea as a group undoubtedly possess the most primitive type of retina, for in them it shows no differentiation except for a gradual increase in the percipient elements towards the central region. The absence of cones may, however, be accepted as a structural adaptation to a nocturnal mode of life. If this is so, it may be argued that the lemuroid type of retina (so far as it is known in living species) could hardly have formed a stage in the evolution of the diurnal retina of higher Primates in which rods and cones are both present.†

In its general structure the retina of *Tarsius* conforms to the lemuroid type. But in the unusual structure of the central point of the retina this prosimian exhibits a line of evolutionary differentiation quite peculiar to itself among Primates. In this specialization, together with the absence of cones, *Tarsius* contrasts strongly with the typical condition in the Anthropoidea. While the tarsioid retina may readily be conceived as the end-result of the progressive specialization of the lemuroid nocturnal type of retina, it can in no way be regarded as a structural precursor of the type characteristic of the higher Primates.

* Hence the term " primordium maculæ " is not very apt.

† The small size of the orbits in certain fossil lemuroids (*e.g.*, *Adapis* and *Notharctus*) suggests that these forms had not yet adopted a nocturnal habitat. Thus it is possible that they possessed a retina of the diurnal type. On the other hand, in all fossil tarsioids of which the skull is at all known (with the exception of *Cænopithecus*) the orbits had evidently attained to proportions comparable with those of the modern *Tarsius*.

In the monkeys, apes and Man, the retina differs from that of all other mammals in the possession of a macula and fovea, with the significant exception of *Nyctipithecus*. The complete absence of this elaboration must probably be regarded as a truly primitive feature, for it seems unlikely that a foveal differentiation, once it had developed in association with enhanced visual acuity, would completely disappear.* Even in the nocturnal tarsier there is a local differentiation in this region, albeit of a different nature. Thus there is some reason to believe—from the evidence provided by *Nyctipithecus*—that the evolutionary appearance of the Anthropoidea occurred at a period when the characteristic differentiation of the anthropoid macula had not yet appeared. From this inference the possibility further arises that the development of the fovea may have occurred independently in the Old and New World monkeys after they had diverged from a common ancestral stock. In the marmoset the foveal excavation is still, as it were, incomplete. In certain monkeys, on the other hand (*e.g.*, *Cerocebus*), the differentiation of the macular region has even surpassed that of the human retina. This observation is of interest in connection with the fact (as recorded in a previous chapter) that in some monkeys the lamination of the main lower visual centre of the fore-brain and the differentiation of the visual cortex may also be more pronounced than is the case in Man.

On the evidence of retinal structure, then, it may be inferred that in the ancestral form common to the Primates as a whole the retina was of a generalized mammalian type, possessing both rods and cones (and thus adapted for diurnal vision)

---

* It might be anticipated that the external ear in *Nyctipithecus* would show some elaboration in its structure as in other nocturnal Primates, but this is not the case (Pocock). On the contrary, the external ear in this genus is rather small and very unobtrusive during life (hence the original name given to this genus—*Aotus*). From this it may perhaps be argued that the nocturnal retina of *Nyctipithecus* was secondarily acquired definitely after the precursors of the Platyrrhines had attained to a pithecoid status.

and showing no macular differentiation. In all the Lemuroidea and *Tarsius* a nocturnal mode of life has led to the disappearance of cones and the formation of a rod-retina, and in *Tarsius* to a still further specialization of quite an aberrant nature. The Anthropoidea presumably separated from the lemuroid and tarsioid stocks before the latter assumed nocturnal habits, for they have preserved the diurnal type of retina. *Nyctipithecus* represents a genus in which the retina has undergone changes similar to those seen in lemurs in association with the adoption of a nocturnal life. Lastly, it is probable that in the early representatives of the Anthropoidea the macula and fovea were as yet undeveloped, and it is not unlikely, indeed, that this sensitive spot in the retina became differentiated independently in the Platyrrhine and Catarrhine stocks.

In the higher Primates the eyes undergo a rotation during embryological development, so that they come to look directly forwards with the optic axes parallel. This position of the eyes—which is reflected also in the plane of the orbital aperture in the skull—is associated with the capacity for stereoscopic vision. It allows of an almost complete overlap of the fields of vision of both eyes, and enables a point in the field of vision to be focussed on corresponding points of both retinæ simultaneously. Herein the Anthropoidea stand in contrast with many lower mammals in which the eyes are placed rather on the lateral surface of the head so that the visual fields do not overlap to any significant extent and each eye receives a different image.

The extreme importance of the acquisition of the capacity for stereoscopic vision in the phylogenetic emergence of the higher Primates has been emphasized by Elliot Smith. We would accordingly refer to his published books and papers on the evolution of Man for an exposition of the educative value of this refinement of the sense of sight in the cultivation of muscular skill and of the incentive which it provides for the evolutionary development of all parts of the brain, especially of those parts underlying the higher intellectual faculties.

The rotation forwards of the eyes is clearly correlated with the reduction of the snout region. In the lower Primates (*Tarsius* and lemurs), therefore, the plane of the orbital aperture is directed laterally to varying degrees, and does not face directly forward as in the Anthropoidea.

In Man stereoscopic vision depends upon or is facilitated by a number of structural features in the visual apparatus besides the position of the eyes. These are the differentiation of the fovea in the retina (which has been discussed above), an incomplete decussation of the retinal fibres in the optic chiasma (whereas in sub-Primate mammals the decussation is practically complete), the delamination of the lateral geniculate body (a terminal visual centre in the brain) into six layers, and the increasing complexity of the nucleus of the oculomotor nerve which controls most of the small extrinsic muscles of the eyeball.

The incomplete decussation of the optic chiasma allows of the representation of functionally corresponding parts of the retinæ of both eyes in the same part of the brain. In the Anthropoidea generally, approximately half of the optic fibres remain uncrossed and thus terminate in the same side of the brain. In *Tarsius* the extent of the decussation is not certainly known, but Woollard's observations suggest that it is almost complete. As regards the Lemuroidea, the only available evidence on this point refers to a brain of the small mouse-lemur (*Microcebus*) which I had the opportunity of studying, and in which one eye had been destroyed by an accident some months before death. From this case I was able to affirm, both by an examination of the fresh brain and of stained sections of the optic tracts, that the decussation is very incomplete—as in the Anthropoidea.[3]

The six-layered type of lateral geniculate body is commonly found in all the Anthropoidea and is sometimes seen also in the Lemuroidea. It has been demonstrated in Man and the macaque monkey (and thus is presumably the case in other Anthropoidea) that alternate layers of each nucleus receive crossed and uncrossed optic fibres respectively. Thus it is interesting to note that in lower mammals such as the dog and cat, in which the optic decussation is practically complete, the geniculate body contains only three well-defined laminæ. It is significant that in *Tarsius* also only three laminæ are found—a fact which is in harmony with Woollard's study of the optic chiasma.

Lastly, we may refer very briefly to the configuration of the oculomotor nucleus. In the Anthropoidea this nucleus becomes complicated by the differentiation of a median group of cells which has been called the nucleus of convergence, and which is functionally related to the converging movement of the eyes essential for true stereoscopic vision. In both *Tarsius* and the Lemuroidea this nucleus of convergence cannot be detected as a well-defined or circumscribed nuclear element.[2]

In summary it may be noted, therefore, that only in the Anthropoidea is the structural basis of stereoscopic vision fully developed. In the Lemuroidea it is foreshadowed in the incompleteness of the optic decussation (at least in *Microcebus*) and the complexity of the geniculate body, while in *Tarsius* even these features are but feebly expressed.

The Ear

In the higher Primates the dominance of the sense of vision is apparently correlated with retrograde changes affecting not only the olfactory but also the auditory sense (if one may judge by the structural modifications of the external ear). In lower Primates, however, the external ear retains the primitive elaboration which is found in most other mammals, for the reason that a nocturnal mode of life necessarily demands considerable auditory acuity.

In the Lemuroidea the external ear is often relatively large.

Pocock's researches show that it finds its simplest and most primitive expression in *Lemur* and *Cheirogaleus*.[9] In these forms it is not very conspicuous (see Fig. 62, A). The helix is a strong ridge, the tragus and antitragus are small and unobtrusive, while the antihelix is continued forward above the external auditory meatus as a simple horizontal ridge—the supratragus or plica principalis. In other forms—*e.g.*, *Microcebus*, *Galago*—the pinna is much larger, is marked by a number of grooves, and can be folded down. Further, the supratragus exhibits a specialization in being expanded into a conspicuous flap. In *Nycticebus* and *Perodicticus* this develop-

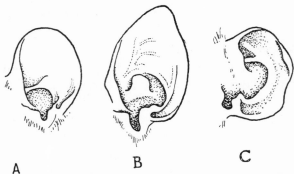

A   B   C

Fig. 62.—The External Ear of A, *Chirogaleus* × ⁴⁄₃ (after Pocock); B, *Tarsius* × ⁴⁄₃; and C, *Cebus* × ²⁄₉ (after Pocock).

ment of the supratragus is also present, while in these forms the tragus and antitragus are rather vestigial processes. In *Chiromys* the pinna is large, but does not show the specializations seen in *Galago*. Its lower portion, just beyond the meatus, is " elevated as in Carnivora, Ruminants, and many other mammals " (Pocock). Nayak has noted that in some genera of the Lorisinæ the antihelix bifurcates into two branches, as in monkeys, the lower of which is the supratragus.

In *Tarsius* the external ear is a highly specialized structure (Fig. 62, B). The pinna is large and grooved, the tragus is a prominent rounded eminence, and the supratragus forms a conspicuous flap which is larger than that of *Galago* and resembles the same structure in certain bats. Woollard's studies indicate that the sense of hearing is very highly developed in *Tarsius*, for the cochlear nuclei and their connections in the brain stem are exceptionally large, and the audito-sensory cortex of the brain is better defined than in lemurs.

In all the Anthropoidea the external ear is relatively small and shows but little variation in the different groups. The pinna is rounded in outline. Its upper edge is always folded in to some extent, and in some Platyrrhines (*e.g.*, *Alouatta*) the entire margin may be inrolled as in Man. The supratragus is always a simple ridge (see Fig. 62, C). Among the Old World monkeys the upper free margin of the pinna in macaques and baboons is produced into a definite point, a primitive feature which is commonly absent in the Platyrrhines. The mobility of the pinna in the higher Primates is much restricted, but it is interesting to note that even in Man there are quite a number of intrinsic and extrinsic ear muscles which provide very convincing evidence for a derivation from a type in which the ear was freely mobile, as in lower Primates and in mammals generally.

The comparative anatomy of the small ossicles of the middle ear, malleus, incus and stapes, merit special attention in the consideration of taxonomic values, because, being sheltered within the petrous bone and serving in all mammals precisely the same function of conducting the vibrations of the drum to the inner ear, they are probably little influenced by the vagaries of the environment. In some animals they may attain to most peculiar proportions, for which it has been impossible to assign a functional reason—*e.g.*, the golden mole (Forster Cooper). Many years ago these ossicles were studied in a large number of mammals by Doran.[5] According to this observer, in the Lemuroidea they much resemble those of the Hapalidæ and *Cebus*. " In the marmoset the crura of the stapes are often fused for some distance below the head, as in mammals of very low grade." This statement is of special significance in deciding the question whether the Hapalidæ may be regarded as truly primitive members of the Platyrrhini. The Catarrhine monkeys differ from the anthropoid apes and Man and approximate to the Platyrrhines and most other mammals " in the straight and little-divergent crura of the stapes." As regards the higher Primates, " in their ossicula, but most markedly in the stapes," the anthropoid apes " are much more allies to *Homo* than to the lower monkeys." Lastly, Woollard states that the ossicles of *Tarsius* resemble those of the lemurs.

The Tongue

Since the epithelial covering of the tongue contains scattered in it the peripheral organs of taste—the taste-buds—it may be conveniently considered among the special senses. But it is important to realize that the tongue subserves also functions such as those associated with deglutition and with the toilet of the teeth, for these uses may be reflected conspicuously in its structure.

In the Lemuroidea the tongue is rather slender and pointed. The knob-like fungiform papillæ tend to become somewhat aggregated towards the tip and are also fairly numerous towards the back of the tongue, while in the intermediate region they are sparse. The circumvallate papillæ are commonly three in number, arranged in the form of an equilateral triangle. This is a primitive mammalian disposition. In the Lemuridæ, according to Sonntag,[10] they are more numerous, reaching to as many as nine, and are typically arranged in a Y formation. A characteristic feature of the dorsum of the lemuroid tongue is the conspicuous development of large thorny conical papillæ on the surface of the pharyngeal region behind the level of the circumvallate papillæ. These conical papillæ are most obtrusive in the Lemuriformes, but they are also present in the Lorisiformes, extending down towards the epiglottis, and they represent a lemuroid specialization which recalls strongly the characteristic appearance of this part of the tongue in Carnivora (Fig. 63, A).

On the under surface of the tongue in all lemuroids, attached along its median axis but free at its tip and lateral margins, is a remarkably well-developed sublingua. This structure is to be distinguished from the frenal lamellæ in the floor of the mouth, which have on occasion been confused with it. The sublingua is regarded by some authorities as essentially a primitive formation—the morphological equivalent of the tongue in lower vertebrates. It is commonly found in marsupials and is also to be seen in the tree-shrews. Vestiges of the sublingua are also present in higher Primates, including

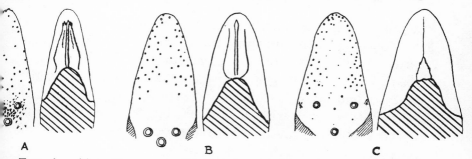

A   B   C

Fig. 63.—The Upper and Lower Aspects of the Tongue of A, *Hemigalago* × 2; B, *Tarsius* × 2; and C, *Hapale* × 1½.

Man. In the Lemuroidea, however, it has attained a high degree of specialization. It forms a thin mobile plate of a horny consistency and presents on its under surface a median axial ridge, the lytta of the sublingua. The apex is always markedly denticulated, and, as in *Hemigalago* (Fig. 63, A), the lateral margins and the lytta may also be finely serrated and there may be a secondary serrated fold on the under surface. The pronounced development of the sublingua with its apical denticles and its free mobility is apparently associated with the peculiar specialization of the lower incisors and canines in the Lemuroidea. It has already been pointed out that these fine styliform teeth are disposed in a closely set pectinate arrangement and are, in fact, actually used for combing the fur. Observation has shown that they are kept clean by the use of the denticulated sublingua, which thus functions as a toothbrush (Wood Jones, Pocock). In the Aye-aye the typical lemuroid modification of the front teeth is not present ; in this animal the sublingua is not free at its apex nor is it denticulated, but, on the other hand, the tip of the lytta is produced into a hook-like structure which is evidently adapted to cleaning the interval between the rodent-like lower incisors. It may be noted that in the Lorisiformes and Lemuriformes the sublingua has become specialized in exactly the same way. If, however, there is a direct relation between this type of specialization and the conformation of the teeth in the incisor region (as seems to be the case), it is certain that the denticulated and mobile sublingua must have developed independently in these two groups. For in the early representatives of the Lemuriformes (*Notharctus* and *Adapis*) the lower incisor and canine teeth do not show the typical lemuroid specialization. That is to say, this dental specialization (and with it, presumably, the specialization of the sublingua) must have arisen after the separation of the lemuriform and lorisiform stocks. Thus this peculiar modification of the sublingua may be regarded as a very definite and fundamental lemuroid tendency, which, if at first latent, finds ultimate expression during the terminal phases of the divergent evolution of this group.

In the tongue of *Tarsius* (Fig. 63, B) the circumvallate papillæ are three in number and arranged in the form of a triangle.* The fungiform papillæ are limited to the anterior two-thirds of the buccal portion of the tongue, not being apparent in the posterior third. The conical papillæ are quite fine and rather elongated hair-like processes, and there are no large thorny papillæ behind the circumvallate papillæ as in the lemurs. This part of the tongue, indeed, appears

---

* Woollard has stated that circumvallate papillæ are absent in *Tarsius*. It must be pointed out, however, that these papillæ may not be easy to define unless the tongue is in a good state of preservation.

macroscopically to be smooth and non-papillated. In this respect the tongue of *Tarsius* is more primitive than that of the Lemuroidea. The sublingua is of simple form, rather fleshy in consistency, adherent at the tip and only slightly free at its lateral margins, and is not serrated. The median lytta is well marked and ends anteriorly in a free point which is slightly bulbous. In many features the tarsioid sublingua approximates to the generalized marsupial structure, and it does not in any way show the specialization characteristic of the lemurs.

The tongue of the marmoset (which probably represents the most primitive expression of this organ in the Anthropoidea) is similar in general form to that of *Tarsius*. It shows the same primitive arrangement of the circumvallate papillæ and the absence of any evident conical papillæ on the surface of the pharyngeal part of the tongue (Fig. 63, C). The fungiform papillæ are rather concentrated toward the tip of the tongue, which may thus present a finely tuberculated appearance. On the under surface is a small triangular sublingua with serrated margins which are free to a limited extent. In other Platyrrhine monkeys the sublingua is represented only by a vestigial fold of mucous membrane—the plica fimbriata—as is also the case generally in the anthropoid apes. In the Catarrhine monkeys, according to Wood Jones,[12] the plica fimbriata is altogether absent, and he rightly regards this as a specialized feature.

Viewing the evidence as a whole, it is clear enough that the tongue in the Lemuroidea—in respect of the sublingua and the development of large conical papillæ in the pharyngeal part—is far more specialized than in the higher Primates. The tongue typical of the primitive members of the Anthropoidea could not therefore be derived in the evolutionary sense from the type characteristic of the recent lemuroids. On the other hand, it seems probable that the tarsioid tongue may readily represent a generalized condition (reminiscent of the marsupials) from which both the lemuroid and the anthropoid tongues have been developed. In the one case this has been accompanied by a marked specialization of the sublingua, and in the other by a marked reduction or complete disappearance.

References

1. BOYD, J. D. : The Classification of the Upper Lip in Mammals. Journ. of Anat., vol. lxvii., 1933.
2. CLARK, W. E. LE GROS : The Mammalian Oculomotor Nucleus. Journ. of Anat., vol. lx., 1926.
3. CLARK, W. E. LE GROS : The Brain of Microcebus. Proc. Zool. Soc., 1931.
4. CLARK, W. E. LE GROS : The Lateral Geniculate Body. Brit. Journ. of Ophthalm., vol. xvi., 1932.
5. DORAN, A. H. G. : Morphology of the Mammalian Ossicula Auditus. Trans. Linn. Soc., vol. i., 1879.
6. ELLIOT SMITH, G. : Essays on the Evolution of Man. London, 1927.
7. KOLLMANN, M., and PAPIN, L. : Études sur les Lémuriens. Archives de Morphologie, vol. xxii., 1925.
8. PAULLI, S. : Ueber die Pneumaticität des Schädels bei den Säugethieren. Morph. Jahrb., vol. xxviii., 1899.
9. POCOCK, R. I. : On the External Characters of the Lemurs and Tarsius. Proc. Zool. Soc., 1918.
10. SONNTAG, C. F. : The Comparative Anatomy of the Tongues of the Mammalia. Lemuroidea and Tarsioidea. Proc. Zool. Soc., 1921.
11. WISLOCKI, G. B. : A Study of the Scent Glands in the Marmosets. Journ. of Mammalogy, vol. ii., 1930.
12. WOOD JONES, F. : The Sublingua and the Plica Fimbriata. Journ. of Anatomy, vol. lii., 1918.
13. WOOLLARD, H. H. : Notes on the Retina, etc. Brain, vol. xlix., 1926.
14. WOOLLARD, H. H. : The Retina of Primates. Proc. Zool. Soc., 1927.

CHAPTER VIII

THE EVIDENCE OF THE DIGESTIVE SYSTEM

THE alimentary tract in the Primates is on the whole rather generalized in form, showing little tendency to any extreme variation. It might be anticipated that the variations which do occur in this system would bear an intimate relation to the nature of the diet, and thus be of little value for assessing affinities. Curiously enough, however, those who have made a special study of the matter have arrived at a very different conclusion. Chalmers Mitchell[3]—in his monograph on the intestinal tract of mammals—makes the statement that " in the mammalian gut, however great may be adaptive resemblances, the inherited element dominates the structure." He shows that in spite of the similarity of the diet in herbivorous marsupials and herbivorous eutherian mammals, there is a profound difference in the gut patterns. On the other hand, the terrestrial Carnivora " display a pattern of intestinal tract essentially similar " and yet " almost every kind of diet is found amongst them—purely carnivorous, piscivorous, omnivorous, frugivorous." This author makes a similar comment on the cæcum, for he concludes that " only in a most general sense can there be said to be a correlation between diet and the presence, length and capacity of the cæcum. There are many exceptions to any general statement, and it seems as if ancestral history were at least as potent a factor as actual diet." Within the limits of the Primates, also, there are examples which support these statements. In this connection it may be noted that the variations of significance which are found in the alimentary canal among the Primates mainly affect the lower end of the tract—the colon—a part of the gut which is not so much concerned with the active digestion of food as with functions of excretion.

The stomach is quite simple in the primitive representatives of each sub-order of the Primates, as in Man himself. It can be subdivided (but not sharply) into fundus, body and pyloric part, and in all respects is similar to the stomach of a generalized mammal. In some of the larger monkeys—*e.g.*, *Semnopithecus* and (to a lesser extent) *Colobus*—the stomach is unusu-

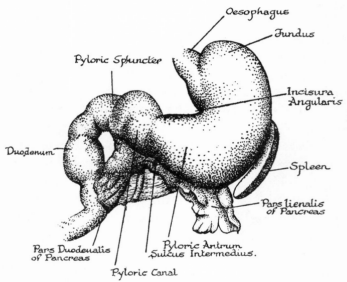

FIG. 64.—THE STOMACH AND ADJACENT ORGANS OF *Tarsius*, TO ILLUSTRATE THE APPEARANCE OF THESE STRUCTURES IN PRIMITIVE PRIMATES. (Woollard, *P.Z.S.*, 1925.)

ally specialized, being very large and elaborately sacculated. This modification may be associated with a peculiar vegetarian diet, and it is also possibly correlated with the absence or reduction of cheek-pouches in these monkeys. It has been suggested that the human stomach is secondarily simplified from such a sacculated organ on the ground that in early fœtal stages the mucous membrane of the fundus is richly corrugated and there is also some indication of sacculation in the region of the lesser curvature. But it is not legitimate to draw phylogenetic inferences from such observations, for during embryological development the mucous membrane of the greater part of the alimentary tract is at first relatively thick and subsequently becomes thinned out by a process of absorption. In the course of this absorption irregularities are produced temporarily which may bear a superficial resemblance to the sacculi of the *Semnopithecus* type. They can hardly be regarded, therefore, as of any evolutionary significance.

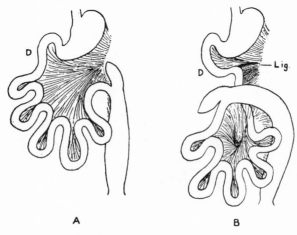

FIG. 65.—DIAGRAM TO ILLUSTRATE THE FIXATION OF THE DUODENAL PART, D, OF THE SMALL INTESTINE BY THE CAVO-DUODENAL LIGAMENT, Lig, WHICH OCCURS IN MOST PRIMATES.

In A, the primitive mammalian disposition of the intestine and mesentery is depicted. In B, the intestinal tract has undergone a rotation as the result of the fixation of the duodenum.

The small intestine is uniform in its general conformation throughout the Primates. Primitively its whole length, including the duodenum, is suspended freely by a median dorsal mesentery. This simple arrangement persists in *Tarsius*, so that in its peritoneal relations the intestinal tract of this small Primate has been compared with that of reptilia and amphibia.[12] In *Chrysothrix* a similar condition is also found. In other Primates (as, indeed, in most mammals) the duodenal part of the small intestine becomes anchored down to the posterior abdominal wall by a band of peritoneum called the cavo-duodenal ligament (Klaatsch). This fixation of the duodenum appears to be one of the factors which conditions or is associated with a rotation of the intestine so that the ileocolic junction comes to be placed in front of the commencement of the small intestine (see Fig. 65). This rotation, moreover, leads to a complication of the mesentery of the proximal part

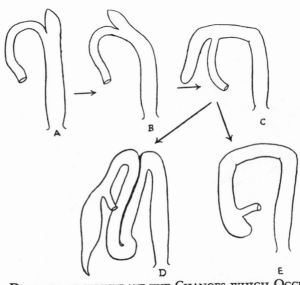

FIG. 66.—DIAGRAM TO ILLUSTRATE THE CHANGES WHICH OCCUR IN THE EVOLUTIONARY DEVELOPMENT OF THE LARGE INTESTINE.

A, The condition in *Ptilocercus*; B, *Tupaia*; C, *Tarsius*; D, Lemuroidea; E, Anthropoidea.

of the colon. In all Primates, with the exception of the large apes and Man (and perhaps certain monkeys), the duodenum is suspended by a definite mesentery or mesoduodenum.

The colon exhibits some instructive modifications which are especially pertinent to the problem of Primate relationships. In order to appreciate the significance of these variations, it is convenient to refer to the main trend of development of this portion of the gut in generalized lower mammals and

Primates as it is diagrammatically represented in Fig. 66. Primitively it seems certain that the mammalian colon was a simple straight tube running back directly to the anal canal and suspended by a dorsal mesentery. At the junction of the small intestine with the colon a diverticulum is found which is called the cæcum. Such a type of colon is characteristic of the reptilia and is found unmodified in the pen-tailed tree-shrew[5] and in certain bats (Fig. 66, A). In the lesser tree-shrew (*Tupaia minor*) the upper end of the colon is inclined to the right, thus producing an incipient division into a transverse colon and a descending colon (Fig. 66, B). This displacement of the ileocolic junction is usually carried further in *Tarsius*, and in this form a distinct transverse colon may be present. In the Anthropoidea the ileocolic region with the cæcum migrates down towards the right iliac part of the abdominal cavity, and thus there now appears a division of the colon into ascending, transverse, and descending components (Fig. 66, E). A significant exception, however, is found in the squirrel monkey (*Chrysothrix*). In this small Platyrrhine the large intestine is very short and is only slightly curved to the right, so that it presents no more than a rudimentary transverse colon and thus exhibits a distinctly tarsioid appearance.

In the Lemuroidea a marked specialization is present which contrasts strongly with the simpler condition of other Primates. In this sub-order the colon is characteristically bent sharply upon itself to form a long and narrow loop (Fig. 66, D). This colic loop, moreover, may become elaborately twisted into a spiral form (*e.g.*, in *Galago*, *Loris*, *Nycticebus*, *Indris* and *Perodicticus*). In the development of such a loop, either in its simple or elaborated form, the lemurs parallel the condition found in certain lower mammals such as Ungulates. It is difficult to ascribe this modification to diet, for it is found equally in the lorises, true lemurs, galagoes and *Chiromys*, among which the diet is by no means uniform. In *Chirogaleus* and *Microcebus*—it is important to note—the lemuroid loop is not present, and thus a more primitive condition is preserved which is not unlike that of *Tarsius*, except that the descending colon is not quite so straight. Now, it is reasonable to assume that the early lemuroids possessed a colon which was at least as simple as that of *Microcebus*, for there is no evidence to indicate that in this genus the colon has become secondarily simplified by a process of reduction. It may be inferred, therefore, that the differentiation of a colic loop occurred after the Lemuroidea had separated from the other Primates in the course of their phylogeny, and that it probably occurred independently in the Lemuriformes and Lorisiformes after these groups had commenced their divergent evolution. This manifestation of a similar evolutionary trend in both these lemuroid groups (not apparently in response to any environmental need) leads to the supposition that they may have been derived from an ancestral form in which this trend was already implicit, even though not at that stage expressed in somatic development.

The characteristic lemuroid tendency to the elaboration of the large intestine suggests quite strongly that the Lemuroidea are rather far removed from the Tarsioidea and Anthropoidea, and certainly militates against the conception that the latter groups have a lemuroid ancestry. It has, however, been claimed that the simple colon of the higher Primates may have been secondarily derived from a more complicated condition, but the arguments for such a conclusion are not convincing. Klaatsch[8] described and figured a small kink in the transverse colon of a six-centimetre marmoset embryo (*Hapale albicollis*) which is not apparent in the adult. He interpreted this as a remnant of the colic loop of lemurs and regarded it as evidence of a lemuroid stage in the evolution of the monkeys. It is probably not justifiable, however, to argue thus from the disposition of the intestine in the embryo, for during fœtal life the arrangement of the gut is modified considerably by the disproportionate size of surrounding viscera such as the liver, and by the fact that, during a considerable phase of development, much of the intestinal tract is extruded through the umbilicus and only in the later stages withdrawn into the abdominal cavity to take up its final position. These temporary displacements and adjustments of the intestine (associated directly with the mechanics of onto-genetic development) probably account in full for the arrangements of the colic bloodvessels in the human adult and the relative elongation of certain parts of the colon in the human infant, features which have both been used to support the thesis that the simplicity of the human gut is of a secondary nature.*

Taking the evidence of the gut pattern as a whole, it is clear that the Lemuroidea—as a group—are more specialized than the other Primates. Indeed, were a comparative mor-phologist to confine his attention to this part of the alimentary tract (to the exclusion of other anatomical evidence), he would legitimately be led to infer that the lemuroid had been derived from the tarsioid type of colon, rather than the reverse.

A cæcum is found in the great majority of mammals. Various opinions have been held as to whether this usual condition is derived from a primitive mammalian form with a single and simple cæcal pouch, with paired cæca (such as is found in birds and, among mammals, in the edentates and Hyrax), or with no cæcum at all. Fortunately we are not here called upon to discuss this vexed question. For in all members of the Primates a cæcal pouch of some kind is present, and, whatever may be the case with mammalia in general, it can hardly be doubted that the Primates themselves have been derived from an ancestral form in which a simple and single cæcum was already present (probably entirely similar in general configuration to the cæcum of the tree-shrews).

In the Lemuroidea the cæcum is relatively large (Fig. 66, D).

* It is to be remarked that some observers regard the hepatic and splenic flexures of the human colon as representatives of colic loops. But they bear no close resemblance to the typical lemuroid " ansa " and may not, therefore, be taken as serious evidence for a lemuroid stage in the phylogeny of Man.

It may be sacculated—as in *Perodicticus, Galago* and *Loris*— or simple and globular—as in *Chirogale* (Wood Jones). Further, in some genera (*Lemur, Galago, Perodicticus* and *Chiromys*) there is a well-differentiated and conical appendix which in macroscopic appearance quite closely approximates to the vermiform appendix of Man and the anthropoid apes. There is no appendix in *Chirogale* or *Loris*.

In *Tarsius* the cæcum is simple in shape but unusually long—almost equalling in length the whole of the colon (Fig. 66, C). It is not sacculated and is devoid of an appendicular process. Johnston[7] notes that in the features of its

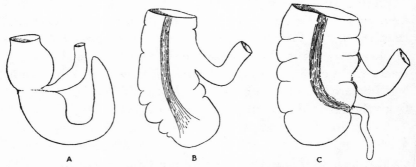

FIG. 67.—THE ILEO-CÆCAL REGION OF A, *Callicebus* (after T. B. Johnston); B, *Macacus*; AND C, MAN.

cæcum and the associated cæcocolic valve *Tarsius* resembles the Platyrrhine monkey *Callicebus personatus*.

In the New World monkeys the cæcum is conical and relatively voluminous, tapering gradually to a pointed extremity which may be elongated and resemble a vermiform appendix (Fig. 67, A). However, the work of Johnston indicates that in this group there is no real differentiated appendix with a characteristic local accumulation of lymphoid tissue; the tapering process can only be regarded, therefore, as the pointed end of the cæcal pouch proper. The cæcum is not sacculated in the Platyrrhines, nor are there any definite muscular bands (tæniæ coli) in its walls. On the other hand, it possesses a well-developed cæcocolic valve.

In the Old World monkeys the cæcum is generally smaller and terminates in a blunt, rounded extremity (Fig. 67, B). Its wall is marked by sacculi and by tæniæ coli, while the cæco-colic valve is absent. It is generally accepted that the vermiform appendix is completely absent in this group. Wood Jones, however, figures the ileo-cæcal region of *Cercopithecus tantalus* in which the appendix is apparently quite well formed, but this would appear to be an exceptional case.

In the anthropoid apes and Man the cæcum parallels that of the Catarrhine monkeys in its relative size, in the presence of definite tæniæ coli and sacculations, and in the absence of a cæcocolic valve. It differs markedly, however, in the possession of a narrow tubular appendix (Fig. 67, C). The significance of the vermiform appendix is still quite obscure, but, in view of its rich blood supply and its histological differentiation, it is almost certainly correct to regard it rather as a specialized (and not a degenerate) structure. Thus, in this regard, the cæcum of the anthropoid apes and Man is probably less primitive than that of the monkeys. Wood Jones, however, has hinted at the conception that the presence of a vermiform appendix is a primitive mammalian feature, since it is found in one of the marsupials—the wombat (*Phascolomys*). But in the vast majority of mammals which, from their general structure, are commonly held to be primitive forms, the appendix is not developed, and it therefore seems much more probable that it represents a specialization which has developed independently in the wombat and some of the Primates. This interpretation is rendered somewhat more feasible by the suggestion of certain authorities that the " vermiform process " of *Phascolomys* is not strictly homologous with the appendix of Man and the apes.

The evidence of the anatomy of the ileocæcal region may perhaps be summarized in the following tentative generalizations. *Tarsius* has retained a cæcum of a primitive type, but manifests a specialization in the direction of progressive elongation. The condition in *Chirogale* indicates that the Lemuroidea have evolved from a form with a simple cæcum, but as a group they show a tendency to develop a more elaborate and sacculated structure with the differentiation of an appendix, and thus parallel the evolutionary changes which occur in higher Primates. In the New World monkeys the cæcum preserves a form which is considerably more primitive than in many lemurs, while in Old World monkeys, although it may be relatively smaller, it shows a higher degree of specialization in the development of sacculations. Finally, in the anthropoid apes and Man a differentiated vermiform appendix appears. Hence there is nothing in this evidence which is incompatible with the thesis that the anthropoid apes and Man—during their phylogenetic development—have passed through a Platyrrhine and a Catarrhine stage and were originally derived from a tarsioid form. It suggests, however, that the Lemuroidea—after their divergence from a common ancestral Primate stock—may have undergone an evolutionary differentiation parallel with, but independent of, the Anthropoidea.

The comparative anatomy of the solid viscera associated with the alimentary tract provides only equivocal evidence for the interrelationships of the Primates, for these structures show a great variability in different members of the same group. The human anatomist is well aware how readily the form of these viscera may be affected and distorted by temporary changes in the hollow viscera which surround them, and that their precise contour, therefore, may vary considerably from individual to individual, and even in the same individual at different times. It is of some interest, however, to record briefly observations which have been made on the liver (Ruge,[10] Duckworth,[6] etc.).

In all Primates with the exception of Man and the anthropoid apes,* the liver is suspended from the diaphragm and the abdominal wall by a complete mesentery, so that no " bare area " devoid of peritoneum is developed. The Lemuroidea,

---

* In *Semnopithecus* the abnormal size of the stomach compresses the liver against the diaphragm, to which it becomes adherent.

on the whole, show an approximation to lower mammals in their tendency to the division of the liver into a relatively large number of lobules.   But this tendency is strictly limited, and it remains questionable whether multilobulation is really a primitive mammalian feature.   A similar tendency, for instance, is manifested in the gorilla, and the liver of this genus thereby stands in marked contrast with that of other anthropoid apes.   The fundamental and primitive subdivision of the liver which is characteristic of the Lemuroidea is shown in *Perodicticus* and *Loris*, in which there are three main lobes— right, central, and left.   Of these, the first is the smallest and the last the largest, while the central lobe is divided into two parts by the umbilical fissure.   In *Lemur* the œsophageal notch is deep (a primitive feature), the caudate lobe is large and closely related to the right kidney, while the papillary lobe is quite small.   Further, the left lateral lobe is extensive and reaches over toward the right margin of the liver. According to Nayak,[9] the liver of *Galago* shows a close approximation to that of *Tarsius*.   In the latter the liver is not well lobulated, and the œsophageal notch is shallower than in many lemurs.   The papillary process is also better developed, fitting into the lesser curvature of the stomach, while the caudate lobe is not so elongated and is partially fused with the right lobe of the liver.   On the other hand, the left lobe is still the largest lobe of the liver in *Tarsius*, and extends well to the right as in the Lemuroidea.

In the marmoset the liver is of quite a simple type, and Beattie states that its lobes and fissures are closely similar to those of *Nycticebus*.[1]   The lobus centralis is deeply cleft by the umbilical fissure into right and left portions as in lower mammals, and these again are separated by deep fissures from the right and left lateral lobes.   The right lateral is now the largest of the lobes of the liver, but the caudate lobe is quite extensive, reaching down to make contact with the right kidney and thus corresponding with the typical lemuroid condition.   In the other monkeys less primitive features may be present. Thus the œsophageal notch is quite shallow in the baboons (though relatively deep in the Cercopitheques), the caudate lobe is reduced in the Catarrhines, and the right lobe shows a tendency towards a relative enlargement so that it ultimately surpasses to a considerable degree the left lobe.

Taking a general view, it may be stated that the liver in the most primitive representatives of the Lemuroidea, Tarsioidea and Anthropoidea closely conforms to the primitive mammalian type. That of *Tarsius* shows a rather more advanced character in the reduction of the caudate lobe, while the incipient reduction of the left lateral lobe, which becomes more pronounced in the higher Primates, is already indicated in the marmoset.

References

1. BEATTIE, J. : The Anatomy of the Common Marmoset. Proc. Zool. Soc., 1927.
2. BEDDARD, F. E. : On the Anatomy of Antechinomys and Some Other Marsupials, etc. Proc. Zool. Soc., 1908.
3. CHALMERS MITCHELL, P. : On the Intestinal Tract of Mammals. Trans. Zool. Soc., 1905.
4. CHALMERS MITCHELL, P. : Further Observations on the Intestinal Tract of Mammals. Proc. Zool Soc., 1916.
5. CLARK, W. E. LE GROS : The Anatomy of the Pen-tailed Tree-shrew. Proc. Zool. Soc., 1926.
6. DUCKWORTH, W. H. L. : Morphology and Anthropology. Cambridge, 1915.
7. JOHNSTON, T. B. : The Ileo-cæcal Region of Callicebus Personatus. Journ. of Anatomy, vol. liv., 1920.
8. KLAATSCH, H. : Zur Morphologie der Mesenterialbildungen am Darmkanal der Wirbelthiere. Morph. Jahrb., 1892.
9. NAYAK, U. V. : A Comparative Study of the Lorisinæ and the Galaginæ. Unpublished thesis.
10. RUGE, G. : Die äussern Formverhaltnisse der Leber bei dem Primaten. Morph. Jahrb., vol. xxix.-xxx., 1902.
11. WOOD JONES, F. : Man's Place among the Mammals. London, 1929.
12. WOOLLARD, H. H. : The Anatomy of Tarsius Spectrum. Proc. Zool. Soc., 1925.

CHAPTER IX

THE EVIDENCE OF THE REPRODUCTIVE SYSTEM

CONSIDERING the uniformity of the functions of the reproductive system, it may be anticipated that its structural components will be much less liable to modification in association with accidental variations of the environment than those of other anatomical systems. But even here secondary adaptations may occur. Thus, for example, the length of the genital tract in the female bears a relation to the length of the symphysis pubis, which itself is partly dependent upon the type of habitat, and the penis may become modified in conformity.

In connection with the male genitalia, it may be noted that in all Primates the testicles descend into scrotal sacs. These sacs are covered by a hairy skin (except in *Lemur catta* in which, according to Pocock,* the scrotum is always naked). It is commonly stated that this descent of the testicles is permanent in the adult, but Nayak[6] records that it is seasonal in *Perodicticus* and also, probably, in *Chiromys*. The same is the case with *Loris*.[11] This is a primitive mammalian feature which has been lost by the vast majority of the Primates. It is interesting to note, however, that the inguinal canals (linking up the sac in which the testicle lies with the abdominal cavity) remain permanently open in *Hapale* and in certain Cebidæ.† The scrotal sacs are situated behind the base of the penis with

---

* The same author states that in the gibbon there is no trace of a scrotum, the testicles in this case descending into the inguinal region in front of the base of the penis. Pocock explains the absence of a pendulous scrotum in this ape as an adaptation to its specialized arboreal activities.

† G. B. Wislocki has recently recorded (*Anat. Rec.*, vol. lvii., 1933) that in the macaque monkey the testes do not descend permanently into the scrotum until the third to fifth year. In the chimpanzee permanent descent occurs as a rule before birth (as in Man).

the exception of *Perodicticus*, in which they occupy a para-penial position, approximating in this respect to the disposition in *Tupaia* and *Didelphys* (Nayak), and representing a primitive condition.

The penis in the Lemuroidea is considerably more specialized than in the higher Primates.* The epithelial covering of the glans is smooth or but slightly wrinkled in *Nycticebus*, *Perodicticus*, *Arctocebus* and *Chiromys*. In all other lemuroid genera it is beset with small or large recurved spicules (" grappling spurs ") disposed in various degrees of complexity, and quite similar to the spinous development on the penis of many lower mammals (*e.g.*, rodents). In *Galago* the spicules may be elaborated into a bidentate or tridentate form. In all the Lemuroidea a bony skeleton is found in the penis, the baculum, which is a common mammalian possession. The baculum varies in its length and contour and, to some degree, in its relation to the urethral canal. In the Indrisine lemurs and in *Cheirogaleus* it forks distally to form two branches, while in the Lemuridæ, the Lorisiformes and *Chiromys* it terminates in a single median thickening.

In *Tarsius* the penis is of a simple form and thus more primitive than in the great majority of lemurs (see Fig. 80). The glans is relatively long and unexpanded, and is covered by a smooth or slightly granular epithelium. The lips of the urethral orifice form rather markedly prominent folds.

In the Anthropoidea the penis is also simple in appearance. In the Platyrrhine monkeys the glans is of a regular ovate or hemispherical form, and only in *Cebus*, it appears, is there a small baculum present, which is limited to the terminal portion of the penis.† In the Old World monkeys (with the

---

* Much of the information on the external genitalia recorded in this chapter has been provided by Pocock's researches on the external characters of the Primates. [7, 8, 9.]

† G. B. Wislocki (*Contr. to Embry.*, No. 133, 1930, Carn. Inst. of Washington) reports that in the spider monkey (*Ateles*) the penis is covered with small cornified denticles which are, however, not more than 0·5 mm. long.

exception of *Cynopithecus*, in which it is absent) there is a small baculum in the glans extending back for a short distance into the body of the penis. The glans is usually rounded or oval and often somewhat expanded, and with the exception of one aberrant species of *Macacus* (*M. speciosa*) it is quite short and truncated. The margin of the glans is commonly marked by a median dorsal or by lateral notches which may be rather deep.

In the anthropoid apes the glans penis in the gibbon, orang and chimpanzee is slender and pointed, showing no expansion as in the monkeys. In the gorilla, on the contrary, it forms a flattened and rounded expansion with a very well defined margin or corona. In these features the gorilla approaches much more closely to the monkeys, and also to Man, than do the other Anthropomorpha. It seems probable that this form of penis, with an expanded and rounded glans which is distinctly demarcated from the body of the penis, is a generalized condition, for it is also seen in the primitive tree-shrew *Ptilocercus*.[2]

In review it may be suggested that the evidence of penile structure indicates that the Lemuroidea are divergently specialized to a considerable degree from other Primates. The Catarrhine monkeys also show a specialization in the development of notches in the corona glandis. The Platyrrhine monkeys, however, retain a generalized (and probably primitive) structure much more comparable with that of Man and the gorilla.

The accessory organs of the male genital system offer little evidence pertinent to the problem of Primate interrelationships. The duct of the testicle (vas deferens) and the duct of the seminal vesicle commonly unite to form a single orifice on each side, which opens into the urethral tract. This is the case in *Tarsius* and the Anthropoidea, and also in *Galago* and *Propithecus* among the lemurs. In *Loris*, *Lemur*, *Perodicticus* and *Arctocebus* a slight specialization is indicated by the fact that these ducts open separately. *Chiromys* further emphasizes its aberrant position by a complete absence of the seminal vesicles.

The prostate gland is faintly bilobular in *Perodicticus, Loris, Galago* and *Chiromys*, the lobes being fused together dorsally to the urethra. In *Tarsius* the gland forms a compact single-lobed structure and lies entirely dorsal to the urethral canal. In the marmoset, according to Beattie's statement,[1] the prostate is identical in its anatomy with that of *Tarsius* and is presumably, therefore, entirely retro-urethral. Sonntag[10] states that this disposition is also to be found in the chimpanzee and the orang. In Man, as is well known, the prostate gland completely encircles the urethra, a condition which is also present in the primitive *Ptilocercus* (in which, indeed, the gland bears a remarkably human appearance).

The female external genitalia of the Lemuroidea—like the male organs of this group—show a marked tendency towards specialization. The clitoris is usually very long and is often pendulous. In *Hemigalago* it may be supported by a baculum. The most remarkable feature, perhaps, is that in all the Lorisi-formes the clitoris is traversed by the urethral canal which opens at its tip. In the Lemuriformes, on the contrary, the usual primitive condition is retained, the urethral orifice being situated at the base of the clitoris (though in the genus *Lemur* it may reach towards the tip). In *Tarsius* the clitoris is quite small and is concealed by the prominent labia minora. In the New World monkeys the clitoris may be short and concealed (*Hapale, Callicebus*) or long and pendulous (*Saimiris, Cebus, Ateles*). In the Old World monkeys there is a similar variation, for while in most representatives of this group the clitoris is elongated and exposed, in the *Semnopithecinæ* and in *Theropithecus* it is quite short and concealed by the labial folds. In the gibbon, according to Pocock, the clitoris is usually short, but in one species (*Hylobates concolor*) it is exceptionally long. These variations are thus of little systematic value.

As regards the internal female genitalia, it may be noted that in all the Lemuroidea the uterus is bicornuate—that is to say, it has a short body and two lateral horns into which the oviducts enter.   This is a primitive condition which is common to placental mammals generally.   In *Tarsius* the uterus is also bicornuate, in this case the horns being relatively short (5 mm. in length) as compared with the corpus uteri (which is 8 mm. long).[14]   All the Anthropoidea possess a uterus simplex, consisting of a single undivided corpus and having no lateral horns.   This type of uterus represents a more advanced morphological condition.   It is interesting to note that the uterus of *Hapale* exhibits certain primitive characters, for the fundus is produced laterally into the tubal region, and on its anterior surface is a median groove which marks off these lateral processes in a way which suggests that they represent lateral horns the fusion of which has not been perfected to the extent seen in other Anthropoidea.[1]

It is convenient in this section to mention the ischial callosities which are so characteristic of the Catarrhine monkeys. These callosities or sitting pads often form very conspicuous naked swellings in the perineal or ischial region (especially in the baboons and mandrills), and in the female they may become enlarged and congested at a certain phase of the menstrual cycle.   Thus they appear to function as secondary sexual characters.   Wood Jones[13] has legitimately emphasized the fact that ischial callosities represent a distinctive Catarrhine specialization.   In the gibbon they are also present, but relatively small.   The other anthropoid apes have no callosities; moreover, there is no evidence that they have ever passed through an evolutionary stage in which callosities were developed.   In this respect, therefore, they are certainly more primitive than the recent Cynomorpha.

One of the most important anatomical studies which provides significant evidence for the interpretation of phylogenetic problems, and which may conveniently be considered under the heading of the reproductive system, is that concerned with the structure and mode of formation of the placenta. When the developing ovum reaches the uterine cavity in placental mammals, it throws out certain membranes—the fœtal membranes—which establish contact with the mucous membrane of the uterus. The apposition or union of these two structures results in the formation of an organ called the placenta, by means of which the embryo is brought into more or less intimate relation with the maternal tissues, and which thereby provides for its nutrition, respiration and excretion during uterine life.

Of the fœtal membranes the most fundamental is the chorion, which completely invests the embryo. The chorion is covered by a layer of epithelium, the trophoblast, and is lined by embryonic connective tissue or mesoderm which contains fœtal bloodvessels. Its surface becomes soon covered with villous processes, the chorionic villi, which penetrate into the uterine mucosa. In the simplest type of placenta these villi merely interlock with corresponding crypts formed in the epithelial lining of the uterus. Such a type of placenta is termed epithelio-chorial, and in this case the maternal blood is separated from the fœtal tissues of the chorion by the endothelial lining of the uterine bloodvessels, the connective tissue of the uterine mucous membrane, and the uterine epithelium. A more advanced type is the syndesmo-chorial, in which the uterine mucous membrane is partly denuded of its epithelium so that this barrier between maternal and fœtal blood is removed. A further stage is represented by the endothelio-chorial placenta, in which the trophoblastic covering of the chorion comes into direct contact with the vascular endothelium of the uterine capillaries (as seen in Carnivora, Cheiroptera and some Insectivora). Lastly, the most efficient type of placenta from the point of view of the intimacy of the relation between fœtal and maternal circulations and the rapid interchange of material between the two is represented by the hæmochorial placenta. In this type the trophoblastic epithelium erodes its way right into the blood sinuses of the uterine wall, opening up large blood spaces filled with maternal blood with which it lies in direct contact without even the intervention of the endothelial lining of the maternal vessels.

There can be little doubt that the epithelio-chorial placenta is the most primitive of these types. This is the view held by most comparative embryologists, including Professor J. P. Hill who is one of the foremost authorities on the subject.* Grosser has pointed out that in the reptilian stages of mammalian phylogeny " the junction of a complete mucosa uteri with the chorion must have started placentation " and that, indeed, " this type of placenta is still found among recent viviparous reptiles."[3] Thus it may be assumed that in the earliest mammals the placenta was of the epithelio-chorial type. The hæmochorial placenta has presumably been derived from such a type during the course of mammalian evolution by the increasing invasive activity of the trophoblast, which shows a progressive tendency to erode the uterine mucosa until finally it reaches the lumen of the maternal blood sinuses.

The placentation of all the Lemuroidea without exception is of the epithelio-chorial type. Moreover, it shows a primitive feature in being diffuse, and not localized as in higher Primates. In its structure the lemuroid placenta manifests " an extraordinarily close general resemblance to the placenta of such an Ungulate as the pig " (Hill), and contrasts very strongly with *Tarsius* and the Anthropoidea in which the hæmochorial structure is found. It has been claimed from time to time (*e.g.*, by Hubrecht,[5] Wislocki,[12] etc.) that the lemuroid placenta has really been derived from a more complicated type by a process of secondary simplification. The arguments on which this claim is based rest partly on the *a priori* assumption that the Lemuroidea have been derived from an ancestral stock represented by *recent* Insectivora in which the placenta is hæmochorial, and they are bolstered up by an appeal

* The phylogenetic significance of the placentation of the Primates has recently been discussed in a fine monograph by J. P. Hill. The short discussion which follows in this chapter is based almost entirely on it.[4]

to the various specializations of lemuroid anatomy, the inference being that since lemurs are rather highly specialized mammals, it would be incongruous to find in them a generalized type of placentation. In the first case, there is no reason to suppose that the modern Insectivora may not have acquired a hæmochorial placenta independently of the Primates, even supposing these two mammalian Orders have been derived from a common ancestral group of primitive insectivores. There are plenty of examples of parallelism of a similar degree within the limits of the Primates (as we have indicated in other chapters). In the second place, specializations in one anatomical system are not necessarily correlated with specializations in other systems. This is a most dangerous form of argument, for the very reverse is often the case. Indeed, it may be taken as quite a general rule that a high grade of specialization affecting one part of the body is usually associated with the retention of remarkably primitive traits in another part. This association of advanced with protean characters in the same animal is seen again and again. We may therefore agree with Hill that there is not the slightest evidence that the simplicity of the lemuroid placenta is not truly primitive, and moreover it is extremely difficult to conceive the steps by which such a reversal of evolution (as would be involved in the derivation of an epithelio-chorial from a hæmochorial placenta) might occur. Hence it can be affirmed that the Lemuroidea retain the ancestral eutherian type of placentation with very little modification, and this certainly harmonizes with the inference drawn from other anatomical and from palæontological data that the sub-order separated from the other Primates at an extremely early stage. But it is interesting to note that in some features in the development of the placenta the Lemuroidea do foreshadow to a slight but significant extent features which are very distinctive of the higher Primates. These are the rapid and early differentiation of the chorion as a whole, and its precocious vascularization from the bloodvessels which develop in relation to the allantois (an outgrowth from the hind-gut) instead of from the bloodvessels of the yolk sac.

In *Tarsius* the placenta is of the hæmochorial type and resembles that of the higher Primates in some points so closely that Hubrecht was led to infer that this small animal should properly be classified as a member of the Anthropoidea. The placenta is a localized discoidal structure and thus contrasts with the diffuse lemuroid type. In its development it approximates to that of the Anthropoidea in the very early differentiation of the chorion and in the way in which it becomes vascularized. Instead of a vesicular allantois such as is found in lemurs and lower mammals, an almost solid mesodermal connecting stalk is formed by which the embryo is attached to the placenta. This connecting stalk is of the same nature as that found in all higher Primates including Man, and probably represents the result of a precocious proliferation of the mesodermal covering of the reduced allantoic diverticulum. By its development, the vascularization of the chorion is enabled to take place very early so that the establishment of a functioning placenta can be rapidly completed, for bloodvessels appear in the body stalk before they are even apparent in the yolk sac or the embryo itself. Clearly this acceleration is of considerable advantage for the nutrition of the embryo in its earliest developmental stages.

In the minute structure of the placenta Hill has shown that there are unusual features in the histological characters of the trophoblast and in its differentiation which render *Tarsius* rather unique among other Primates. Indeed, on these grounds, he considers that " the *Tarsius* placenta is too specialized to have been the actual forerunner of that of the Pithecoids," that it " has evolved along lines of its own, and that such general resemblances as it presents to the Platyrrhine organ is the result of developmental parallelism."

But he concedes the possibility that the Anthropoidea may have arisen from some other branch of the tarsioid stock in which these structural peculiarities (which are, it seems, by no means of a fundamental nature and doubtless merely generic specializations) had not occurred. In the formation of its yolk sac *Tarsius* is also somewhat aberrant among the Primates, for this structure is relatively large and is apparently endowed with important absorptive properties. Lastly, it may be noted that although the placenta of *Tarsius* evinces such a remarkable approximation to that of the higher Primates, it shows one or two features in which it is more primitive, and reminiscent, so to speak, of the generalized mammalian type, such as the mode of formation of the amniotic sac from folds which later coalesce.

In the Anthropoidea the early developmental processes concerned in the establishment of the placenta are still further accelerated, and a very definite advance on the tarsioid condition is manifested in the formation of the amniotic sac which no longer arises from folds but appears (as it were spontaneously) in the middle of the embryonal ectoderm. The chorion is formed more precociously than in *Tarsius*, and, as in the latter, the vesicular allantois is replaced by a connecting stalk in which fœtal bloodvessels appear at a very early stage. In the monkeys, with extremely few exceptions, two separate discoidal placentas are formed. It is interesting to note that *Hapale* is quite unique among the higher Primates (including *Tarsius*) in normally producing twins and sometimes triplets. This feature serves to emphasize the primitive status of the marmoset, for plural gestation is certainly a primitive mammalian attribute.

The Platyrrhines and Catarrhines show certain differences in their placentation which are the result mainly of the profuseness of the trophoblastic proliferation in the former, and in this character the New World monkeys approximate more to *Tarsius*, while the Old World monkeys are more progressive. Professor J. P. Hill observes that by comparison " the development of the Platyrrhine placenta is a slow and cumbrous process, involving so much time that it only reaches a condition of what we may call structural efficiency at quite a late period in gestation," while " the Catarrhine, on the other hand, by abbreviation and acceleration of the developmental processes, has speeded up the development of its villous placenta in the most remarkable way." Finally, he points out that " the Catarrhine placenta is capable of carrying on its full functions immediately the fœtal circulation is established, and it does so not only much earlier, but in what would seem to be a much more efficient manner (judging from structural relations) than is the case in the Platyrrhine."

The placentation in the anthropoid apes and Man is very similar to that of the Catarrhine monkeys, in which, indeed, its distinctive characters are already clearly foreshadowed. The early embryo elaborates its placenta in a still more precocious manner and gains a most intimate relation with the maternal tissues by very rapidly burrowing its way (by means of its trophoblastic activity) through the uterine epithelium and into the wall of the uterus. Thus it differs notably from the lower Primates in which development takes place in the lumen of the uterus. Unlike the monkeys also, only one discoidal placenta is formed.

In summarizing the evidence of the placentation, we may emphasize the retention by the Lemuroidea of extremely primitive characters which closely approximate to those which are postulated for the basal ancestral eutherian stock, but repeat that their placental development also presents certain modifications which adumbrate faintly the distinctive features of the higher Primates. In other words, the placentation of this sub-order of the Primates " provides just the requisite basis for the manifestation of those adaptive modifications in the developmental processes which characterize in such distinctive fashion the evolutionary history of the higher groups of this order " (Hill).

In its early development and placentation *Tarsius* occupies a position between the Lemuroidea and the Anthropoidea, but is evidently very considerably nearer to the latter. This provides one of the most convincing lines of evidence that the Anthropoidea and *Tarsius* have arisen from a common ancestral form. We may remark that if such a form already possessed features of placental structure common to the modern monkeys and *Tarsius*, these features must be of very ancient origin, for other anatomical evidence (already reviewed) indicates that the Anthropoidea certainly separated from the tarsioid stem quite close to the point at which the latter became differentiated from the basal Primate stock.

According to Professor Hill's interpretation, the structure of the placenta points to the fact that the Platyrrhines and Catarrhines arose from a common precursor which had attained to a pithecoid status, and that a diphyletic origin is extremely unlikely for these two groups. On the other hand, they probably separated very early from the parent stem and underwent a measure of parallel evolutionary development. Lastly, the Catarrhine stock provided the basis for the phylogenetic emergence of the anthropoid apes and Man.

References

1. BEATTIE, J. : The Anatomy of the Common Marmoset. Proc. Zool. Soc., 1927.
2. CLARK, W. E. LE GROS : The Anatomy of the Pen-tailed Tree-shrew. Proc. Zool. Soc., 1926.
3. Grosser, O. : Human and Comparative Placentation. Lancet, May, 1933.
4. HILL, J. P. : The Developmental History of the Primates. Phil. Trans. Roy. Soc., vol. ccxxi., 1932.
5. HUBRECHT, A. A. W. : The Descent of the Primates. New York, 1897.
6. NAYAK, U. V. : A Comparative Study of the Lorisinæ and the Galaginæ. Unpublished thesis.
7. POCOCK, R. I. : On the External Characters of the Lemurs and Tarsius. Proc. Zool. Soc., 1918.
8. POCOCK, R. I. : On the External Characters of the South American Monkeys. Proc. Zool. Soc., 1920.
9. POCOCK, R. I. : The External Characters of the Catarrhine Monkeys and Apes. Proc. Zool. Soc., 1926.
10. SONNTAG, C. F. : The Morphology of the Apes and Man. London, 1924.

11. SUBBA RAU, A., and HIRIYANNAIYA, S. : The Urogenital System of Loris. Journal of Mysore University, vol. iv., 1930.

12. WISLOCKI, G. B. : Carnegie Contributions to Embryology, vol. xx., 1929.

13. WOOD JONES, F. : Man's Place among the Mammals. London 1927.

14. WOOLLARD, H. H. : The Anatomy of Tarsius Spectrum. Proc. Zool. Soc., 1925.

CHAPTER X

THE RELATION OF THE TREE-SHREWS TO THE PRIMATES

ALL those who have made a study of Primate anatomy and who have considered the problem of the evolutionary origin of the Primates are in agreement that this group of mammals is derived from an ancestral form of essentially insectivorous type. That is to say, the recent Insectivora mark the closest approximation among living mammals to the progenitor of the Primates. In the Order of Insectivores are usually classed the tree-shrews (Tupaioidea*), which in many of their features actually show a remarkable similarity to primitive Primates, so much so, indeed, that some zoologists have deemed it legitimate to remove them from the Insectivora and group them with the Primates. Thus Carlsson[1] as a result of a detailed study suggested that they really represent a sub-order of the " Prosimiæ " ; Gregory[11] classes them with the lemurs, tarsioids and Anthropoidea as a super-order " Archonta " ; and Wood Jones[16] links them with the lemurs under the heading of Strepsirhini, which he contrasts with the Haplorhini (Tarsioidea and Anthropoidea). While the Tupaioidea undoubtedly show Primate affinities, they are also in many ways remarkably primitive mammals, and it seems probable that in their general structure they represent a tolerably close

* The tree-shrews have for many years been associated by systematists with the elephant-shrews (Macroscelidoidea) as the Menotyphla, in contrast to the Lipotyphla which comprise all other genera of recent Insectivora. But the anatomical resemblances between the tree-shrews and the elephant-shrews are few and of questionable significance, while the differences are so marked that there is no doubt that they are really quite widely divergent—and certainly not closely related.[9]

approximation to the earlier phases in the evolution of the Primates. Hence it is of great importance to consider in some detail the anatomy of these small animals.

The living Tupaioidea are represented by the family Tupaiidæ, of which there are two sub-families, Tupaiinæ and Ptilocercinæ. The latter comprises one species only— *Ptilocercus lowii*, the pen-tailed tree-shrew—which has a geographical distribution limited to Borneo, Sumatra and the Malay Peninsula. It is a completely arboreal mammal, constructing nests in hollow branches. The Tupaiinæ have

FIG. 68.—A YOUNG SPECIMEN OF THE LESSER TREE-SHREW (*Tupaia minor*).
From a photograph taken by Mr. Banks. Note the wide abduction of the pollex and hallux.

a wider distribution, extending over India, Burma, the Malay Peninsula, Sumatra, Java and Borneo, and are represented by a number of species. The smaller species (*e.g.*, *Tupaia minor*) are very active arboreal mammals, being found among the higher branches of the trees, while the larger species (*e.g.*, *Tupaia ferruginea*) are rather bush-animals, inhabiting the undergrowth and the lower branches of the trees. Superficially the tree-shrews bear a resemblance to squirrels, for which at a distance they may at first be readily mistaken by the field naturalist. This resemblance is enhanced by the relatively long and bushy tails which many of them possess (Fig. 68).

Recently a fossil tupaioid, discovered in Oligocene deposits in Mongolia, has been described in detail by Simpson.[15] This animal, *Anagale gobiensis*, is a member of another family (Anagalidæ) of the tree-shrews and, as we shall see, offers some remarkable corroborative evidence for the Primate affinities of the Tupaioidea as a whole.

In the sum of its anatomical characters *Ptilocercus* is much

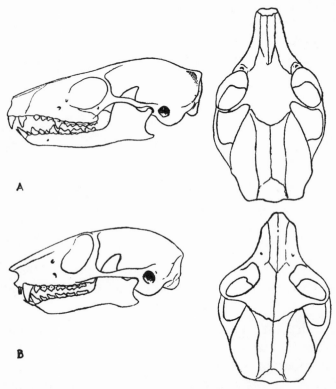

A

B

FIG. 69.—LATERAL AND DORSAL VIEWS OF THE SKULL OF A, *Ptilocercus* × $\frac{4}{3}$, AND B, *Microcebus* × 1.

more primitive than the genus *Tupaia*, while *Anagale* exhibits many features (at least in regard to its dentition and skeletal structure) in which a closer approach is made to the Lemuroidea.

The general appearance of the Tupaiidæ is shown in the accompanying figure of a young specimen of the lesser tree-shrew. It will be observed that the body proportions are of a generalized nature and suggest the nimbleness and agility which are characteristic of these animals among the branches. The eyes are rather large, the snout does not project very conspicuously (at least in the smaller species), the fore-limbs are shorter than the hind-limbs, and the long tail is employed as a " balancer." The furry coat is usually of a neutral colour, though in some of the larger species of tree-shrews it takes on a rufous tint, especially over the tail. The full complement of vibrissæ characteristic of a generalized eutherian mammal is present, including small carpal and calcaneal groups. In *Ptilocercus* the tail, except at the base and the tip, is relatively naked, being covered by a scaly epidermis. It will be convenient to consider the significant anatomical features of the tree-shrews under the headings of the several systems.

Skull.—The remarkable superficial resemblance of the skull of *Ptilocercus* to that of the smaller lemurs is shown in Fig. 69. As in the Lemuroidea, the snout region is not so markedly produced as in many lowly insectivores, the orbit is completely surrounded by a bony ring (which, however, is incomplete in *Anagale*, as it is also in the primitive adapid *Pronycticebus*), the plane of the orbital aperture is rotated forwards to a moderate degree, the malar is perforated by a foramen, the brain-case is relatively expanded (much more so in the Tupaiinæ) and lacks a median sagittal crest, the zygomatic arch is slender, and the mandible is slightly built. But the resemblances are by no means confined to generalities. In many particulars the similarity between the skull structure of the tree-shrews and the lemurs amounts to an actual identity.[5] Thus the articulations of the bony elements which form the orbito-temporal region are the same as those in the Lemuriformes. The orbital plate of the palatine is extensive and reaches forward to gain contact with the lachrymal, separating the frontal from the maxilla. This condition is not found in the other insectivores, for in them there is a wide fronto-maxillary contact. The maxilla just reaches the lachrymal at the inferior orbital margin.

The whole construction of the tympanic region seems to be identical with the lemuriform condition and is paralleled nowhere else among the mammalia. The bulla is relatively large, and the ectotympanic forms a simple ring enclosed within it. The course and arrangement of the arteries in this region are also the same, for the entocarotid pierces the bulla close to its posterior margin and runs up in the medial wall of the tympanic cavity. It gives off a stapedial branch, and in the Tupaiinæ this stapedial artery is relatively large while the

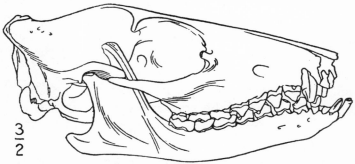

FIG. 70.—LATERAL VIEW OF THE SKULL OF *Anagale gobiensis* × 1½.
(G. G. Simpson, *Amer. Mus. Nov.*, 1931.)

continuation of the entocarotid is markedly reduced (a characteristic feature of the Lemuriformes, in which the brain is vascularized mainly by the vertebral arteries, whereas in other Primates the entocarotid becomes the dominant source of blood-supply). The bulla in the Lemuriformes appears to be constructed of the same elements as that of the tree-shrews, but a little uncertainty exists on this point because the embryological development of the skull in the Malagasy lemurs has not yet been worked out in detail. In the Tupaiidæ there is a small tympanic wing of the alisphenoid which forms part of the roof of the tympanic cavity. This is a primitive or metatherian feature which is also observed among the tarsioids (*e.g., Necrolemur*).

In addition to these cranial features there is in *Ptilocercus* a partial or incipient flexion of the face on the basicranial axis which becomes, of course, greatly exaggerated in the higher Primates. The lachrymal extends to a small degree on to the face in *Ptilocercus*, and to a greater extent in *Tupaia*. In both the lachrymal foramen is situated in the orbital margin. In *Anagale* there is a marked reduction of the premaxilla, again recalling the lemuroid condition. Among the primitive features of the skull may be noted the backward extension of the malar bone to the glenoid fossa, a common marsupial disposition. Lastly, the arrangement and completeness of the various cranial foramina for nerves and vessels in *Tupaia* are similar to those of the lemurs. The expansion of the brain-case in *Tupaia* and especially *Anagale*, even involving to some degree the frontal region, is an important feature, for it is especially characteristic of the Primates.

Limbs.—In the tree-shrews the limbs are on the whole of a very generalized structure, but in some respects they exhibit features which can only be regarded as foreshadowing or even quite closely approaching the characteristic structure of the Primate limbs. In the fore-limb the clavicle is strongly built, and the scapula and humerus closely resemble those of primitive Primates ; it has already been recorded that, according to Gregory, the humerus of *Ptilocercus* is quite similar to that of the fossil *Plesiadapis*. There is a well-developed entepicondylar foramen, and the capitellum is relatively large and separated from the trochlea by a distinct groove. The radius and ulna are free and allow of some degree of pronation and supination. In the manus the generalized eutherian pattern of palmar pads is preserved, and of the digits the third is the longest. The terminal phalanges are provided with sharp compressed claws. In *Anagale* these are more specialized than in recent tree-shrews, for the terminal phalanges are slit to accommodate a ridge on the under aspect of the claw as in certain insectivores and edentates.

The carpus in *Ptilocercus* is generalized, but in *Tupaia* the scaphoid and lunate bones may be fused. The os magnum shows some degree of lateral compression which recalls the characteristic lemuroid condition, but the os centrale is well separated by it from the unciform. The first metacarpal has a concavo-convex facet for articulation with the trapezium, and the thumb is capable of a wide range of movement. It can be powerfully abducted, and, with converging flexion, it can also to some degree be brought into opposition with other digits for grasping purposes.* It is from such a hand that the typical manus of the Primates must certainly have been derived.

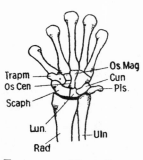

FIG. 71.—THE CARPUS AND METACARPUS OF *Ptilocercus*, PALMAR ASPECT × 2½. (*P.Z.S.*, 1926.)

In the hind-limb the os innominatum, femur, tibia and fibula are of a generalized type, the latter being quite separate, and in some features they show (especially the ilium) an approximation to the primitive Primates. A rather closer approach is made by the stouter fibula of *Anagale*. The plantar pads in *Ptilocercus* are all separate and thus present a generalized pattern. In *Tupaia* the thenar and first interdigital pads are fused together as is usually the case in lemuroids. The third digit is the longest.

* The phrase " opposition of the thumb " is limited by some authorities to the perfected movement by which the thumb can—at the carpometacarpal joint—be completely rotated so that its palmar surface is brought into direct and firm contact with the palmar surfaces of the other digits. It is probable that the ability to perform this movement is only fully developed in Man (largely as the result of the perfection of the neuromuscular mechanism which controls it), and thus it may be argued that the human hand can alone be said to possess an " opposable thumb." Nevertheless, the elements of opposability are to be observed in other Primates to varying degrees, and even in the tree-shrews, though the movement may not have acquired the refinement and completeness which is shown in the human thumb. See Mr. Banks' account of the living pen-tailed tree-shrew quoted in my monograph on this animal for the use which it makes of the pollex in grasping its food.

In recent tree-shrews all the digits are provided with well-developed claws. Certain authorities (*e.g.*, Lorenz)[12] have laid stress on the presence of claws in the tree-shrew and have argued that this feature is sufficient to contradict any suggestion of affinities with the Primates. This curious line of reasoning takes no account of the presence of well-marked claws in

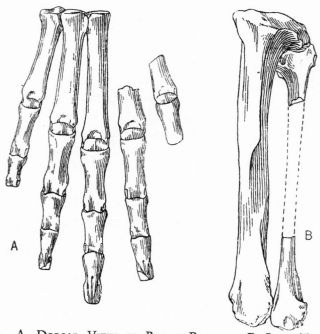

FIG. 72.—A, DORSAL VIEW OF RIGHT PES, AND B, LEFT TIBIA AND FIBULA OF *Anagale gobiensis* × 1½. (G. G. Simpson, *Amer. Mus. Nov.*, 1931.)

Chiromys or on one or two of the pedal digits in the other Prosimiæ. In any case, however, the argument is no longer valid, for Simpson has demonstrated quite clearly that in the fossil tupaioid *Anagale* the terminal phalanges of the foot were spatulate in form (as in primitive lemurs such as *Notharctus*), and the inference is unavoidable that they must have borne flattened nails. Thus the Tupaioidea evince a propensity for replacing claws by true nails, which is a highly distinctive feature of the Primates. The contrast in *Anagale* between the markedly specialized claws of the manus and the flattened nails of the pes is very noteworthy, and it is consonant with the conclusion drawn from the anatomy of living forms (*vide supra*, p. 136) that in the evolutionary origin of the Primates the hinder extremity assumed a typical Primate structure before the fore-limb. Further, it is of unusual interest in so far as it demonstrates that the morphological differences between a true claw and a true nail are not so great as some anatomists would believe—since they may coexist in a well-developed form in the same animal. Hence it would seem to be quite unwise to attempt to draw a sharp distinction between the nailed digits of the Primates and the clawed digits of more primitive mammals.

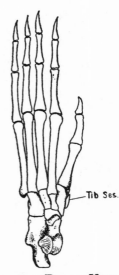

Tib Ses.

FIG. 73.—DORSAL VIEW OF HIND-FOOT OF *Ptilocercus* × 2½. (*P.Z.S.*, 1926.)

The tarsus in the Tupaiidæ preserves a primitive mammalian pattern, and, though in their contour the individual tarsal elements show a number of resemblances to those of the lemurs, they do not manifest specializations such as the extension of the front part of the os calcis or the compression of the mesocuneiform which are characteristic of many lemuroids. The first metatarsal of *Ptilocercus* is stouter and conspicuously shorter than the other metatarsals, and its base is marked by a rounded peroneal tubercle, indicating an incipient specialization of the hallux. The facet on the entocuneiform with which it articulates is, moreover, oblique and somewhat saddle-shaped. Associated with this conformation of the bones, the hallux is capable of wide abduction, and a converging flexion which allows it to some extent to be brought into opposition with the other digits and thus used for grasping purposes.

In general, then, the limbs of the tree-shrews conform remarkably well with those which might be postulated for the ancestral Primates; they present a mechanism of hand and foot from which, by progressive refinement and specialization, the more mobile extremities of recent Primates could readily be derived; and, indeed, they already show in some instances a definite tendency toward the development of nails and an opposability of the hallux and pollex which are such characteristic features of the Primates as a whole.

Dentition.—The dental formula of the Tupaiidæ is $\dfrac{2 \cdot 1 \cdot 3 \cdot 3}{3 \cdot 1 \cdot 3 \cdot 3}$. In *Anagale*, however, the first premolar is retained, and there are strong indications that the third upper incisor was present. Thus this extinct tupaioid shows the generalized eutherian formula which was characteristic of the earliest Primates— $\dfrac{3 \cdot 1 \cdot 4 \cdot 3}{3 \cdot 1 \cdot 4 \cdot 3}$.

The upper incisors in *Tupaia* and *Anagale* are relatively short and styliform, and recall to a slight extent the characteristic reduction of these teeth in modern Lemuroidea. In *Ptilocercus* they are definitely specialized and powerful teeth, the second incisor being rather premolariform. The lower incisors are markedly procumbent, less so in *Anagale* than in *Tupaia* (in which they lie almost horizontal). This procumbency again suggests an incipient lemuroid condition, by no means closely approaching, however, the very specialized arrangement of the lower incisors of recent lemurs. The canines are rather reduced and incisiform in modern Tupaiinæ, and in *Tupaia melanura* and *Ptilocercus* the upper canine has two roots. In *Anagale* the upper canine is single-rooted and stouter than the incisors, while the lower canine is somewhat incisiform. Thus the canines in the tree-shrews show no close resemblance to those of modern lemurs, though in *Anagale* they are not markedly different from the primitive canines of *Notharctus*.

Of the upper premolars, the last tooth (P⁴) shows some degree of molarization in *Tupaia*, bearing on its crown para-, meta-, and mesostyles, and the three cusps of the trigone, as well as possessing three roots. In *Ptilocercus* it is much simpler, for besides a large and pointed paracone it shows only a small protocone (and no metacone or definite cingulum). As in *Tupaia* it is three-rooted. In *Anagale* the last upper premolar is subquadrate with probably an incipient hypocone.

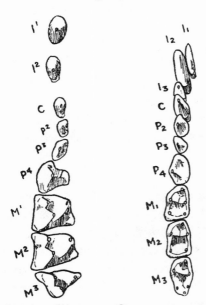

FIG. 74.—OCCLUSAL SURFACE OF THE TEETH OF *Ptilocercus* × 4.

In the lower series also, the last premolar is molariform in *Tupaia*, and less so in *Ptilocercus*. In *Anagale* it has a high trigonid with protoconid, metaconid and a small paraconid, while the talonid is about equal in length to the trigonid.

The upper molars of *Tupaia* are of primitive form except that the styles are rather well developed, the mesostyle being split into two. In the development of these small and sharp cusps on the external cingulum *Tupaia* imitates a common condition in the Insectivora, but that it is a parallel development is indicated by the fact that in the more primitive tupaioid dentition of *Ptilocercus* the mesostyle is absent (though small para- and metastyles are present). It has already been noted that mesostyles are characteristic of the upper molars of *Notharctus*. In *Ptilocercus* there is an incipient true hypocone on the first and second upper molar. The upper molars of *Anagale* are very similar to those of *Ptilocercus* except that the hypocone seems to have been rather better developed, giving a more quadrate form to the crown.

In the lower molars of *Tupaia* and *Ptilocercus* the three cusps of the primitive trigonid are preserved, while the talonid, which is rather short, shows an entoconid and hypoconid. The last molar of *Tupaia* is reduced, but in *Ptilocercus* this is not so and it has quite a distinct hypoconulid. In *Anagale* the worn character of the teeth precludes a definite statement about the paraconid, while the talonid is rather elongated in the last molar and the hypoconulid " is attached to the internal half of the trigonid (as in the primitive lemuroid and tarsioid M_3) " (Simpson).

From this brief account of the dentition, it will be observed that the modern tree-shrews show a number of features in which an approach is made to the dentition of primitive Primates. It may be said, indeed, that—apart from the moderate degree of generic specialization—the tupaioid dentition provides in several points a basis from which the lemuroid dentition may well have been derived. The Oligocene *Anagale* offers still more suggestive evidence of Primate affinities in this connection, for, as Simpson has pointed out, this fossil tupaioid " approaches the primitive Primates more closely than do the recent tupaiids " in the following dental characters :

(*a*) $P\frac{4}{4}$.
(*b*) Premolars more affected by molarization.
(*c*) Upper molars subquadrate.
(*d*) Paracone and metacone less crescentic and more external.
(*e*) Trigonids short and narrower than talonids.
(*f*) Lemuroid projection of hypoconulid on M_3.

Thus it is evident that, on the basis of the dentition, it is not possible to draw a clear-cut distinction between the Tupaioidea and the primitive Primates. This lack of distinction is further emphasized by the difficulty which palæontologists have had in deciding the status of the fossil Plesiadapidæ.

For whereas the late Dr. Matthew, an authority of the widest experience, associated these forms with the tree-shrews under the heading of the Menotyphla, it is now held by many authorities (Simpson, Abel, Stehlin, Teilhard de Chardin), and indeed generally agreed, that they should legitimately be classified in the Order of Primates.

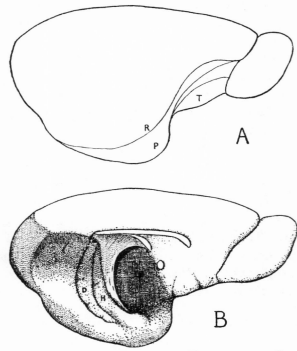

FIG. 75.—A, LATERAL AND B, MEDIAL VIEWS OF THE CEREBRUM OF *Tupaia minor* × 4. (*P.Z.S.*, 1932.)

The Brain.—Seeing that the progressive elaboration of the brain (and especially of the neopallial cortex) is such an outstanding characteristic of all the Primates, it becomes a matter of peculiar interest to study the brain of the Tupaioidea. Recent researches have provided a considerable amount of detailed information on the subject.[3, 9]

The brain of *Tupaia* is found to be relatively large when other sub-Primate mammals of an equivalent size are taken for comparison. In *Tupaia minor* the brain-weight is $\frac{1}{26}$ of the body-weight. In *Microcebus*—an animal which weighs approximately twice as much as *Tupaia minor*—the ratio is $\frac{1}{36}$. The relative volume of the brain of *Tupaia* as compared with insectivores is well emphasized by Fig. 76, in which it is seen that, in cross-section, it is as large as that of *Centetes*, and yet the latter has a body-weight about fourteen times as great as that of the tree-shrew. The olfactory regions of the tupaiid brain all show a marked reduction—the olfactory bulbs are relatively small (though still definitely larger than in recent Primates), the olfactory tubercle is flattened, the hippocampal formation forms quite a narrow strip on the medial surface of the cerebrum, and the piriform cortex has become displaced by the expanding neopallium so that not more of it is to be seen

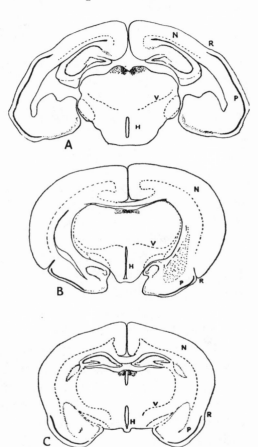

FIG. 76.—TRANSVERSE SECTIONS THROUGH CORRESPONDING LEVELS OF THE BRAIN OF A, *Centetes*; B, *Tupaia minor*; AND C, *Microcebus*. All × 4.

H, Hypothalamus ; N, neopallium ; P, piriform cortex ; R, rhinal fissure ; V, ventral medullary lamina of the thalamus.

Note the resemblance between *Tupaia* and *Microcebus*, and the corresponding contrast with the insectivore brain.

on the lateral aspect of the hemisphere than is the case with *Microcebus* (Fig. 46). The corpus callosum is elongated and straight as in primitive lemuroids, while the fornix and ventral commissure are quite small.

The neopallium shows a relative expansion in the occipital and temporal regions, though the frontal pole remains somewhat attenuated. The temporal lobe projects down to form a definite " pole," and in front of this the hemisphere is excavated by the large orbits to form a broad and shallow Sylvian fossa as in *Tarsius* (but less extensive). A projecting occipital pole is also present which overlaps the whole of the midbrain as well as the anterior surface of the cerebellum. The only neopallial sulcus which is visible macroscopically is a short and shallow precalcarine sulcus ; but microscopically there are also to be seen definite short intercalary and pre-Sylvian sulci. The histology of the neopallium is quite elaborate—the cortical areas are as clearly circumscribed and as richly cellular as those of *Microcebus*. In comparison with the latter the frontal and parietal association areas are less extensive, while the prefrontal cortex is hardly evident (see Fig. 46). The visual cortex of *Tupaia* is remarkably well developed.[4] In extent it covers a relatively large area of the hemisphere, reaching forward and downward on the medial aspect as far as the corpus callosum. In its microscopical structure this cortical area shows a degree of differentiation which, in animals of equivalent size, is only paralleled among the Primates. It contains a double granular layer with a conspicuous intervening stria.

A study of a series of the true Insectivora shows that as the brain increases in size the piriform lobe becomes larger and the hippocampal formation more elaborate, while the neopallium remains relatively small in extent (as, for instance, in *Gymnura*, the largest of the living insectivores). In the case of *Tupaia* the increase in size of the brain is marked by a developmental trend in precisely the contrary direction, for here the neopallium expands and the piriform lobe and hippocampal formation undergo retrogressive changes as in the Primates.

In the thalamus the pulvinar is proportionately as large as in *Microcebus*. The main part of the lateral nucleus is definitely smaller but nevertheless does show some increase as compared with insectivores, and this has led to a characteristic widening of the dorsal surface of the thalamus, which becomes much more marked in the recent Primates. The lateral geniculate body is not only large, but shows an intrinsic structural differentiation which appears to foreshadow the characteristic Primate formation.[8] Its constituent cells show incipient signs of lamination, and the laminæ are disposed vertically. It is from just such a structure that the everted and inverted types of geniculate body in the Primates almost certainly were initially derived (Fig. 47). There is no differentiation of superficial large-celled laminæ such as is found in all living Primates with the solitary exception of *Tarsius*. The elaboration of the lateral geniculate body of *Tupaia* is correlated with the large size of the optic nerve and the high grade of development of the visual cortex.

The mid-brain of *Tupaia* appears remarkably primitive, for the anterior colliculus forms a great rounded prominence which is relatively larger than in any other mammal. Thus it resembles (superficially at least) the optic tectum of the reptilian brain. This excessive development of the anterior colliculus is, again, associated with visual acuity. The posterior colliculus, on the contrary, is rather small. The cerebellum is of a simple mammalian type, but a little more elaborate than that of *Microcebus* or *Tarsius*.

In *Ptilocercus* the brain is much more primitive than that of *Tupaia* and indeed is of a very generalized form. The visual neural mechanisms show no outstanding development (for the pen-tailed tree-shrew is a crepuscular animal), the olfactory regions of the brain are relatively larger and the neopallium less extensive. In many ways this brain provides a transitional stage between the " advanced " type found in

Tupaia and a lowly type such as is seen in some of the Insectivora (*e.g.*, the hedgehog). Judging from the figures of the skull of *Anagale*, it appears that the brain of this fossil form was of the *Tupaia* type, but rather larger than, for instance, in *Tupaia minor*, and perhaps with a slightly better developed temporal pole of the cerebral hemisphere.

That the brain of *Tupaia* exhibits without doubt many suggestive features which are usually regarded as characteristic of primitive Primates is perhaps most clearly expressed and summarized in tabular form. We may affirm that *Tupaia* shows a closer approximation to the Primates than any other insectivore (or, indeed, any other mammal of an equivalent size) in the following points:

1. The relative size of the brain as a whole.
2. The expansion of the neopallium, accompanied by a displacement downwards of the rhinal fissure.
3. The formation of a distinct temporal pole of the neopallium.
4. The backward projection of the occipital pole.
5. The presence of a shallow calcarine sulcus, and of short intercalary and pre-Sylvian sulci.
6. The well-marked lamination and richness in cell-content of the neopallial cortex.
7. The degree of differentiation of the cortical areas in general.
8. The retrogression of olfactory mechanisms as shown by a relative reduction of the olfactory bulb, a flattened olfactory tubercle, a reduced piriform lobe, and a very evident reduction of the hippocampal formation.
9. A pronounced elaboration of the visual apparatus of the brain, and especially of the higher centres. This is evidenced mainly by a high grade of differentiation of the visual cortex and the intrinsic elaboration of the lateral geniculate body.
10. An advanced degree of differentiation of the nuclear elements of the thalamus, among which is especially to be noted the large size of the pulvinar which has by some authorities been regarded as an obtrusive feature of the Primate thalamus.

On the other hand, the brain of the tree-shrew still remains more primitive than that of any other *living* Primate in :

1. The relative size of the brain as a whole.
2. The relative size of the olfactory bulb.
3. The absence of a retrocalcarine fissure.
4. The small extent of the parietal and frontal association areas of the cortex.
5. The low grade of lamination of the lateral geniculate body.
6. The small size of the main part of the lateral nucleus of the thalamus.
7. The large size of the anterior colliculus.

Systematists might be reluctant to admit the tree-shrews to the status of a Primate because in these features their brain has by no means reached such an advanced stage of development as that of the small living Primates. But the brains of extinct Primates such as the Plesiadapidæ (judging from fragmentary remains of the skull) must have been still more primitive than that of *Tupaia*. Thus it hardly appears that the primitive characters of the tupaiid brain are sufficient to offset the advanced or Primate-like characters in the problem of assessing Primate affinities.

The Digestive System.—The stomach in the Tupaiidæ is simple, as in the majority of Primates. The whole length of the intestine is suspended by a free mesentery. The small gut is rotated so that the duodenum lies behind the ileo-cæcal region. The large intestine is a short wide tube which runs straight back to the anal region without the formation of any loops. At the upper end of this portion of the gut is a small conical cæcum. This disposition of the colon is extremely primitive and, indeed, almost reptilian in appearance (Fig. 78). In *Tupaia* the upper end of the colon shows a slight inclination towards the right, suggesting the initiation of a transverse colon (Fig. 66).

It has already been argued (*vide supra*, p. 201) that the simple arrangement of the intestine in the tree-shrews forms the

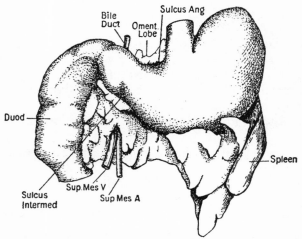

FIG. 77.—THE STOMACH AND ADJACENT VISCERA OF *Ptilocercus* × 3.
(*P.Z.S.*, 1926.)

Compare with Fig. 64, showing the same structures in *Tarsius*.

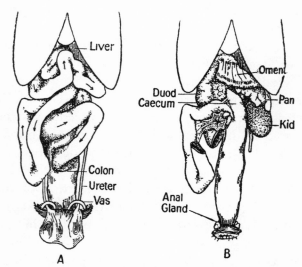

FIG. 78.—THE ABDOMINAL VISCERA OF *Ptilocercus*, A, WITH THE
SMALL INTESTINE IN POSITION, AND B, WITH THE GREATER PART
OF THE SMALL INTESTINE REMOVED × 1. (*P.Z.S.*, 1926.)

basis from which the Primate colon-pattern has been derived, and, in fact, the colon of *Microcebus* is but little more elaborate than that of *Tupaia*.

In the possession of a cæcum, the Tupaioidea differ from all the Insectivora except the elephant-shrews, and approximate to the Primates.*

The Reproductive System.—The external genitalia of the male present rather a striking resemblance to those of the Primates. This is due mainly to the pendulous condition of the penis and to the size and contour of the scrotal sacs

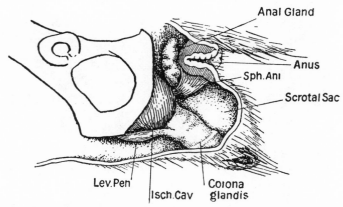

FIG. 79.—THE PENIS AND SCROTAL SAC OF *Ptilocercus* FROM THE LEFT SIDE.

The testicle has been removed from the scrotal sac × 4. (*P.Z.S.*, 1926.)

(Fig. 80). In *Tupaia* the testes are permanently retained in the scrotum as in practically all Primates (*Perodicticus*, *Loris* and *Chiromys* appear to be exceptions), while in *Ptilocercus* the descent of the testes is seasonal. The shape of the penis in *Ptilocercus* is not unlike that of the human organ, being relatively rather short and with an almond-shaped glans. In *Tupaia* it is longer and more attenuated in conformity with the longer symphysis pubis and genital tract in the female of this genus. There is no baculum or os penis in the tree-

* It should be noted, however, that the cæcum is said to be absent in *Tupaia tana*.

shrews, and herein they contrast with all the Lemuroidea. In the Tupaioidea the scrotal sacs occupy a parapenial position as in some lemurs (*e.g., Perodicticus*), and thus to some extent provide a transitional stage between the prepenial position characteristic of marsupials and the postpenial position of most Primates. But, in fact, these apparent variations in the position of the scrotum are not altogether due to a dis-

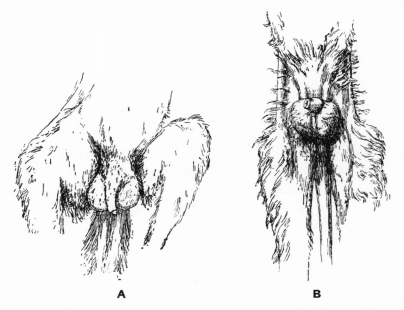

A   B

Fig. 80.—The Male External Genitalia of A, *Tupaia belangeri*, and B, *Tarsius* (Woollard).

placement of this structure itself, but rather to the differing dispositions of the penis.

Of the accessory sexual organs, the prostate gland of *Ptilocercus* is a compact unilobular structure which in macroscopic and microscopic appearance is remarkably similar to that of all the Primates, and thus contrasts with the elaborate and subdivided gland found in such insectivores as the hedgehog. In *Tupaia* the gland appears to be quite small, for what Kaudern has described as a large prostate gland in this genus is almost certainly the vesicular gland. In *Ptilocercus* there are on either side of the prostate two sac-like structures which open by ducts into the urethra. Of these, one is the vesicular diverticulum, which joins the terminal part of the vas deferens and opens by a common orifice with it, while the other is the vesicular gland which opens separately and immediately proximally. A well-developed uterus masculinus is found in both genera of the Tupaiidæ.

The external female genitalia of *Ptilocercus* and *Tupaia* con-

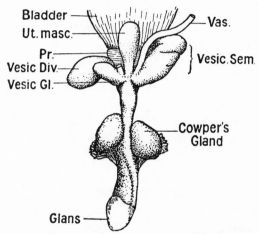

FIG. 81.—THE MALE URETHRA OF *Ptilocercus* VIEWED FROM THE DORSAL ASPECT × 3. (*P.Z.S.*, 1926.)

form closely to the usual arrangement found in primitive Primates. The labia minora are rather prominent, projecting well beyond the less well-defined labia majora. The latter fuse behind to form a median raphe in the perineum. There is a small clitoris, which is grooved on its under surface and completely concealed by the labia.

The internal female genitalia are of a primitive form, the uterus having two relatively long cornua which fuse to form a short body. In the shortness of the corpus uteri the tree-shrews are rather more primitive than the lemurs. The oviducts are of moderate length and slightly coiled. The proportions of the genital tract in the female *Ptilocercus* are shown in the scale drawing in Fig. 83.

The placentation of *Tupaia* is of the hæmochorial type and corresponds to that of the higher Primates. As in the monkeys, the placenta is double and discoidal, while Strahl has noted in its minute structure some resemblances to that of *Tarsius*.

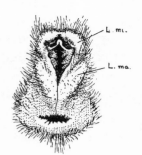

FIG. 82.—THE EXTERNAL FEMALE GENITALIA OF *Ptilocercus* × 3.

L.ma, Labium majus ; L.mi, labium minus.

Special Sense Organs.—In the structure of the rhinarium the tree-shrews belong to the strepsirhine type of Pocock and Wood Jones, and thus show agreement with the lemurs. The olfactory cavities contain an ethmo-turbinal system of a simple character, which corresponds to the general type described by Paulli[14] for the Insectivora and Lemuroidea. That is to say, there are four endo-turbinals including the naso-turbinal, and the second possesses two scrolls. The ecto-turbinals are, however, reduced to two, and here the Tupaiidæ show agreement with the majority of the Lemuroidea and contrast with the Insectivora in which there are three.

The retina in *Tupaia*[17] is of the diurnal type with a preponderance of cones (resembling that of the Anthropoidea but lacking a fovea), while in *Ptilocercus*[6] it is of the nocturnal type—a rod retina—and analogous therefore to the type found in all Lemuroidea and *Tarsius*. The external ear in

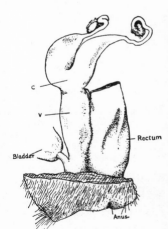

FIG. 83.—THE INTERNAL FEMALE GENITALIA OF *Ptilocercus* × 2.

C, Corpus uteri; V, vagina.

these two genera displays a corresponding contrast. In *Ptilocercus* it is large and well developed, showing a slight elaboration in the production of the antitragus into a flap-like process. In *Tupaia*, on the contrary, the pinna has undergone retrogressive changes, so that in general appearance it bears a remarkably close resemblance to the external ear of the higher Primates (compare Fig. 84, B with Fig. 62, C). Thus in the conformation of the external ear the tree-shrews show the same evolutionary propensities as are manifested in the Primates. In the nocturnal type the ear becomes elaborated, and in the diurnal type it undergoes precisely the same retrogressive modifications which are seen in the Anthropoidea.

The ossicles of the middle ear offer some significant evidence. Doran,[10] writing as far back as 1879, states that the malleus of *Tupaia* " differs from that of any other insectivorous mammal "

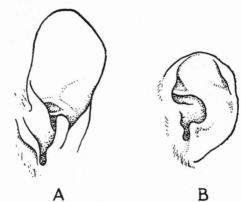

Fig. 84.—The External Ear of A, *Ptilocercus* × 2, and B, *Tupaia ferruginea* × 2.

and, in being neckless and devoid of a laminar process, " much resembles that of some of the lower Primates, especially *Midas* or *Hapale*, or certain lemurs." As regards the incus, " in general characters this bone is very like the same in many monkeys and lemurs."

The tongue of the tree-shrews shows a remarkable lemuroid character in the development of a well-defined sublingua with a serrated margin. It is somewhat better developed in *Tupaia*, in which it is more free. It is supported by a median thickening—the lytta—which, however, does not contain a cartilaginous or muscular support as in the lemurs. There is some slight evidence that the sublingua has undergone reduction in *Tupaia*, for Carlsson found that in a fœtal specimen of *Tupaia javanica* it was relatively longer, reaching nearer to the apex of the tongue. The apex of the tongue shows a rather rich accumulation of fungiform papillæ, which bear witness to the development here of a special sensitivity which Sonntag regarded as a distinctive tendency in the Primates.

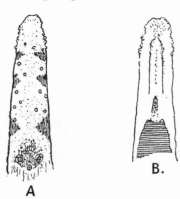

FIG. 85.—A, DORSAL ASPECT AND B, VENTRAL ASPECT OF THE TONGUE OF *Ptilocercus* × 2. (*P.Z.S.*, 1926.)

On the dorsal surface of the tongue are three circumvallate papillæ.

The Muscular System.—The muscular system of the Primates has not been considered in this thesis for (in so far as information is at present available) it provides no unequivocal evidence regarding the precise interrelationship of the various Primate groups. But there are many features in which the myology of *Tupaia* shows resemblances to that of the Prosimiæ and at the same time departs from the Insectivora, and thus it deserves special mention here. The main points may most clearly be expressed by the following list, which indicates the normal conditions found in *Tupaia* and in the Lemuroidea, but absent in the lipotyphlous Insectivora :

1. M. piriformis present.
2. M. caudo-femoralis present.
3. M. brachio-radialis present (but absent in *Ptilocercus*).
4. M. obliquus internus passes mainly dorsal to the rectus muscle. In the insectivores (with the exception of the Talpinæ) it passes ventral.
5. The elements of the quadriceps extensor muscle of the thigh are relatively well differentiated at their insertion.
6. The insertion of M. popliteus does not extend down further than the upper quarter of the tibial shaft.

7. The flexor brevis digitorum takes origin from the deep flexor tendons, the plantar fascia, and also directly from the os calcis.

8. The extensor brevis digitorum provides a tendon for the hallux, and, moreover, in *Ptilocercus* this portion of the muscle is partly differentiated from the main mass to form an extensor brevis hallucis.

9. A well-developed abductor hallucis is present.

10. Differentiation of M. peroneus digiti quarti.

The following conditions are frequently absent among the lipotyphlous insectivores, though found in *Tupaia* and the Lemuroidea:

1. M. teres minor present.
2. M. sartorius present.
3. M. crureus differentiated.
4. M. coraco-brachialis fully developed.
5. Differentiation of the scalene muscles.
6. Coracoid head of the biceps present.
7. Differentiation of M. flexor brevis pollicis and M. flexor brevis minimi digiti.

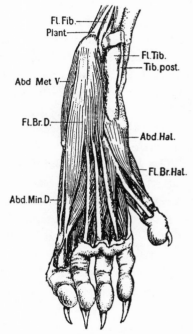

FIG. 86.—THE MUSCULATURE OF THE SOLE OF THE FOOT OF *Ptilocercus* × 3. (*P.Z.S.*, 1926.)

Lastly, in the following points of muscular anatomy *Tupaia* departs from the normal lemuroid structure and displays a more primitive condition:

1. Origin of M. extensor longus digitorum from the femur.
2. M. tenuissimus present (represented, however, in the Primates by the short head of the biceps).
3. Pyramidalis and accessorius present. Although absent in lemurs, these muscles are found in the Anthropoidea, while the accessorius has been reported as an exceptional condition in *Tarsius*.

4. Undifferentiated opponens muscles of the pollex and minimus digits in the manus. These muscles are partially differentiated in the lemurs, but are absent in *Chiromys*.

A general survey of these data indicates that the tree-shrews differ from the Insectivora and show an approximation to the Primates in a much greater differentiation of the limb musculature. This is closely correlated with a greater freedom in the movements of the limbs, and the differentiation of the small muscles of the manus and pes is of peculiar interest in so far as it provides evidence for an enhancement of the independent mobility of the digits, and especially the hallux and pollex. Thus the myological anatomy of the Tupaiidæ corroborates the implications which arise from a study of the limb structure as a whole.

The Affinities of the Tree-shrews.—It has been stated above that the Tupaioidea are commonly classified as members of the order Insectivora. On the other hand, the evidence which has here been reviewed shows that in many features they afford rather marked structural contrasts with insectivores generally, and at the same time display evidence of affinities with the Primates. The diagnostic characters of the order Primates are not easy to define clearly, for, as Wood Jones and others have stressed, many of their outstanding structural characteristics are really the result of the retention of primitive mammalian features. And, as others have pointed out (*e.g.*, Chalmers Mitchell), the retention of the same primitive features in two different forms is not necessarily indicative of a close affinity between them. But the resemblances of the Tupaioidea to the Primates are by no means all of a negative nature (though even these have their value as negative evidence). In some characters the tree-shrews manifest a similarity or an identity with progressive or specialized trends which are highly distinctive of the Primates among all other mammalian Orders. Of these characters may be emphasized again the enlargement of the brain with reduction of the olfactory mechanisms, advanced differentiation of the visual centres and neopallial expansion; the propensity for developing flattened nails instead of claws (*Anagale*); the conformation of the molar teeth; the free range of movement of the pollex and hallux; the differentiation of the small muscles of the manus and pes; the remarkable lemuroid features in the tympanic and orbito-temporal regions of the skull (including the course and branches of the entocarotid artery); the structure of the tongue; the conformation of the small ossicles of the middle ear; and the permanent retention of the testes in the scrotal sac in *Tupaia*.

If the comparative anatomy of the Lemuriformes is considered in detail, it appears certain that this group of Primates in its phylogenetic history—and after diverging from the generalized basal mammalian stock which also gave rise to the other Primate sub-orders—passed through a stage represented by small and primitive mammals which must have resembled so closely in all fundamental details the tree-shrews that no systematist could avoid classing them in the same super-family—Tupaioidea. It may be argued, however, that these resemblances are fortuitous and can be adequately explained by evolutionary convergence or " homoplasy " dependent on similar habits of life. This argument might be valid in regard to some of the more general features, but it can hardly be applied to the detailed similarities. The latter, in many cases, can by no means be related to identical functional requirements of the environment. And even were this possible, it would still not be legitimate to infer a parallelism unless there were coincidently marked structural contrasts which would necessarily indicate a very distant relation between the two groups. Such contrasts do not exist, for the specializations peculiar to members of the Tupaioidea are entirely of a generic nature and are relatively trivial.*

* *Tupaia* shows no resemblance to the lemurs in its hæmochorial type of placenta (that of *Ptilocercus* has not yet been studied). But a hæmochorial placenta is found in *Tarsius* and the higher Primates. On the current interpretation of placental evolution, the placentation of *Tupaia* has presumably been derived from a simple and primitive type such as has been retained by all recent lemurs.

Thus it seems certain that the tree-shrews—although remarkably primitive in many ways—do in fact represent an offshoot from the Primate phylum *after* the latter had become definitely differentiated from other mammalian phyla. It appears, on the anatomical evidence, that we may go even further than this general conception. Wood Jones includes the Tupaioidea in the group Strepsirhini together with the Lemuroidea. But in the skull structure the tree-shrews show such detailed resemblances to the Lemuriformes that it may be deemed more in accordance with the facts to include them in this subdivision. In other words, it is not improbable that the Lemuroidea had already dichotomized into their two main groups, Lemuriformes and Lorisiformes, before they had reached the early phase of evolution represented by the tree-shrews. This conclusion might seem difficult to accept were it not that collateral evidence is available which points in the same direction. The fragmentary remains of the Palæocene and Eocene genus *Pelycodus* show clearly enough that this was an astonishingly primitive form, and yet it is evidently an early representative of the Notharctinæ which (as Gregory has shown) were definitely differentiated as true lemuriforms. Further, if the Plesiadapidæ are correctly interpreted as precursors of *Chiromys*, the conception of a very early separation of the lemuriform and lorisiform stocks is still more strengthened. We may emphasize again that the Plesiadapidæ are now generally agreed to be early Primates, and yet the dental characters of this extinct family are sufficiently similar to those of the Tupaioidea for the late Dr. Matthew to have grouped them together under the Menotyphla.

Systematists may find difficulty in associating the Tupaioidea with the Primates if they consider only the living representatives of these groups, for, with such a limited view, it is clear enough that the former are very much more primitive than the latter. But if fossil forms are taken into consideration, especially with reference to *Anagale*, plesiadapids, and the early Notharctinæ, it is at once evident that such a structural hiatus by no means exists.

This thesis is not deeply concerned with the taxonomic problems of the systematist. It is certain that hard-and-fast lines between different mammalian groups cannot be drawn, and as palæontological evidence becomes more abundant, the difficulties of classification must correspondingly increase. But, with all the evidence at present available, it is suggested that the position of the tree-shrews would be more correctly interpreted by finally removing them from their association with the Insectivora and including them among the Primates, of which, no doubt, they represent extremely primitive forms.

References

1. CARLSSON, A. : Über die Tupaiidæ. Acta Zoologica, vol. iii., 1922.
2. CLARK, W. E. LE GROS : On the Myology of Tupaia minor. Proc. Zool. Soc., 1924.
3. CLARK, W. E. LE GROS : On the Brain of Tupaia minor. Proc. Zool. Soc., 1924.
4. CLARK, W. E. LE GROS : The Visual Cortex of Primates. Journ. Anat., vol. lix., 1925.
5. CLARK, W. E. LE GROS : On the Skull of Tupaia. Proc. Zool. Soc., 1925.
6. CLARK, W. E. LE GROS : The Anatomy of the Pen-tailed Tree-shrew. Proc. Zool. Soc., 1926.
7. CLARK, W. E. LE GROS : On the Tree-shrew. Proc. Zool. Soc., 1927.
8. CLARK, W. E. LE GROS : The Lateral Geniculate Body. Brit. Journ. of Ophthalm., vol. xvi., 1932.
9. CLARK, W. E. LE GROS : The Brain of the Insectivora. Proc. Zool. Soc., 1932.
10. DORAN, A. H. G.: Morphology of the Mammalian Ossicula Auditus. Trans. Linn. Soc., vol. i., 1879.
11. GREGORY, W. K.: The Orders of Mammals. Bull. Amer. Mus. Nat. Hist., vol. xxvii., 1910.
12. LORENZ, G. F.: Ueber Ontogenese und Phylogenese der Tupaiahand. Morph. Jahrb., vol. lviii., 1927.
13. NAYAK, U. V.: A Comparative Study of the Lorisinæ and the Galaginæ. Unpublished thesis.

14. PAULLI, S.: Ueber die Pneumaticität des Schädels bei den Säugethieren.   Morph. Jahrb., vol. xxviii., 1899.

15. SIMPSON, G. G. : A new Insectivore from the Oligocene of Mongolia.   Amer. Mus. Nov., No. 505, 1931.

16. WOOD JONES, F. : Man's Place among the Mammals.   London, 1929.

17. WOOLLARD, H. H. : Notes on the Retina, etc.   Brain, vol. xlix., 1926.

CHAPTER XI

THE EVOLUTIONARY RADIATIONS OF THE PRIMATES

In the foregoing chapters we have considered various anatomical systems of the Primates with the intent to see what
inferences may be drawn from them severally in regard to
the origin and interrelationships of the subdivisions of the
Order. The evidence which has been adduced shows clearly
enough that these subdivisions must have commenced an
evolutionary segregation at a very early time in the history
of the group as a whole, and that, after the point of separation
from a common parent-stem had been passed, they underwent
a considerable degree of parallel development. In the present
chapter the evidence of the various anatomical systems must
be collated and synthesized in order to give a more comprehensive picture of the phylogenetic history of the Primates.

Palæontological investigations have demonstrated that
living mammals represent but a very few of the end-products
of diverse evolutionary trends. These few have survived
throughout the geological ages as the result, no doubt, of
some particular structural advantages which they have
achieved, and through which they have secured a more or
less complete harmony with the environment in which they
find themselves. The vast proportion of mammalian species
and genera has become extinct, and they represent so many
evolutionary experiments which have ultimately met with
no success.

Now, it is necessary to take into full consideration both
recent and fossil forms in order to arrive at a conception of
the evolutionary trends which have determined the direction
of phylogenetic development, and in this way to apprehend

the precise relationship of one group with another. For—as we have already emphasized—the distinctive features of the Primates or of their subdivisions are not to be defined merely by a consideration of the anatomical characters of the few terminal products of evolution which happen to have survived to the present day, but rather by reference to the dominating evolutionary tendencies which have characterized these groups since their first differentiation from a basal and generalized ancestral stock. Only by this method of approach can we assess the possibilities or probabilities of the direct derivation of one group from another. Thus, for example, if a survey of a whole series indicates that there is a prevailing and progressive tendency in that series towards the reduction and loss of a primitive character X, it may be deemed extremely improbable that the series could have given rise, during the later phases of its evolution, to another series in which the character X is retained or evinces a tendency to progressive specialization.

The data brought forward in the previous chapters provide considerable ground for the propriety of dividing the Primates into three sub-orders, Lemuroidea, Tarsioidea and Anthropoidea, which are rather sharply distinguished by their evolutionary tendencies, and each of which adopts specializations of a contrasting nature. There have been put forward from time to time suggestions for an alternative classification of the Primates, and some of these will be noted. For our present purpose, we may adopt the orthodox classification into these three sub-orders, and it will be convenient to consider each of them separately in order to trace the lines of evolutionary development which they have followed.

Lemuroidea

The modern lemurs can be divided into two main groups, which show a strong contrast with each other in the structure of the tympanic and the orbito-temporal regions of the skull, the mode of vascular supply of the brain, the external genitalia (of the female), certain features of the tongue and the nasal cavities, etc. These groups are the Lemuriformes and the Lorisiformes. It has been demonstrated that, in view of the divergent cranial characters, it is extremely improbable that either group stands in an ancestral relationship to the other. On the contrary, it seems that they must both have been derived from quite a primitive form in which these particular features still retained the generalized eutherian condition. This inference is corroborated by the palæontological evidence.

Notharctus, Adapis, Pronycticebus and *Aphanolemur* (all Eocene lemuroids) show by the specialized construction of the tympanic region of their skulls that they were true lemuriforms. Hence it can be positively affirmed that the Lemuriformes and Lorisiformes had already separated from the parent-stem at least by the Eocene period. But it is possible to go further than this. Early members of the Adapidæ (and forerunners of *Notharctus*) are represented by the fossil remains of *Pelycodus*. Though the skull of this genus is not known, it appears to carry the Adapidæ (a family of the Lemuriformes) back to at least basal Eocene times. Lastly, the Plesiadapidæ provide evidence in the same direction. It has been pointed out that certain continental authorities with some reason regard these primitive lemuroids as representatives of the early stages in the evolution of the Aye-aye—a lemuriform which still exists to-day. If this interpretation is correct, the fact that plesiadapids have been found in Palæocene deposits of Europe and America strengthens still more the inference that the Lemuriformes and Lorisiformes had already diverged in Palæocene times.

This conclusion has far-reaching implications. Judging from the dental and skeletal remains of these fossil lemuroids, the common ancestral stock must have been represented by extremely primitive, small-brained mammals, with a generalized dentition and a skull closely resembling that of a very lowly insectivore. Even in Eocene times the Lemuriformes still retained many primitive features in the size of the brain, the conformation of the incisor teeth and canines, the presence of four premolars, the simple structure of the molars, the fronto-maxillary contact in the orbito-temporal region (*Adapis* and *Pronycticebus*), the absence of a bony ring encircling the orbit (*Pronycticebus*), and the arrangement of the carpal and tarsal elements (*Notharctus*). Yet the modern lorisiforms and lemuriforms both present a number of highly specialized features in common, the majority of which (it seems) were not manifested in the early lemuriforms after they had set off on their own evolutionary course. Hence it appears certain the lemuroid traits such as the procumbency of the lower incisors and the canines, the relatively large size of the brain and the fissural pattern of the cerebral cortex, the formation of a bony ring round the orbit, the reduction of the premolars and the elaboration of the molars, and the distortion of the carpal and tarsal bones, developed independently in these two groups *after* their separation from a common point of origin. We have also seen reason to believe that other highly distinctive lemuroid characters such as the specialized sublingua and the elaboration of the colon pattern arose as an expression of parallelism. Thus we are led to infer that these two groups were derived from an ancestral stock in which a tendency to the development of such peculiar features was already implicit even though not yet manifested.

Chiromys is of special interest in this connection. So aberrant is this genus in many ways that some authorities would include it (with the Plesiadapidæ) in a separate sub-order of the Primates—Chiromyoidea. But the essential identity of many of its cranial characters, its colon pattern, carpus, digital characters, placentation, etc., betray its relationship to the lemurs and especially to the lemuriform group. On the other hand, the peculiar features of its dentition, the hypertrophy of its premaxilla, the preservation of primitive characters in the brain, the orbito-temporal region of the skull, olfactory cavities and claws (among other things), do indicate that it represents a group which separated from the main lemuriform stem very early after the emergence of the

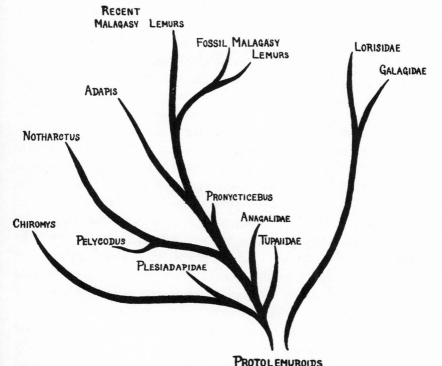

FIG. 87.—SCHEMA ILLUSTRATING THE EVOLUTIONARY RADIATIONS OF THE LEMUROIDEA.

The main morphological trends of evolution are indicated, but not the geological horizons from which the remains of the extinct genera have been derived.*

latter from the basal Primate stock, and that it has almost certainly undergone some degree of parallel development with the other lemuriforms.

* It is hardly possible to employ a diagram of this kind without conveying an impression of finality which is by no means intended. It is urged, therefore, that these schemata should not be taken at their face value, but should be seriously considered only in conjunction with the reservations stated or implied in the text.

Lastly, the tree-shrews add further corroboration to the conception of an early differentiation of the Lemuroidea. Evidence has been brought forward in Chapter X in support of the thesis that the Tupaioidea are really primitive lemuroids, and that, in fact, they are to be regarded as surviving offshoots which early arose from the lemuriform stem. If this is so, it may be inferred that the Lemuriformes and Lorisiformes parted company at least before the evolutionary stage represented by the tree-shrews had been reached.

We may now give a tentative and summary account of the course of evolution of the Lemuroidea as a whole. In Fig. 87 is shown a schematic representation of their phylogenetic relationships so far as they can be deduced from the anatomical and palæontological data at present available. This and the following two schemata represent *morphological* trends in the phylogenetic tree, and the geological horizons at which the various extinct genera have been found are by no means indicated by the end-points of the branches. The diagrams are intended solely to give a general idea of the relative phases in the line of descent at which these genera probably branched off to follow their several evolutionary trends.

The earliest progenitors of the Lemuroidea (which may be designated the Proto-lemuroid stock) must have been extremely generalized mammals which would not be distinguished at first from the basal placental stock (from which they were derived) by any obvious structural features. While still small creatures with small and macrosmatic brains, imperfect ossification of the tympanic region, large olfactory cavities, generalized dentition, orbits not encircled by a bony ring, and a generalized limb structure, they split up into two divergent groups, the Lemuriformes and Lorisiformes. Of the latter, it is very remarkable that nothing is known about their subsequent history up to the present time. No fossil remains which can be certainly attributed to the Lorisiformes are known.* Possibly their evolutionary development took

---

* *Pronycticebus* is an adapid, and *Pseudoloris* a tarsioid.

place in a part of the world whose geological deposits have not yet been explored in detail.

Soon after the emergence and segregation of the lemuriform stock, the anomalous line of evolution which culminated in the Aye-aye must have branched off. If the Plesiadapidæ are early forerunners of *Chiromys*, this separation presumably occurred at a time when the early lemuriforms were at least as primitive as the type represented by the fossil plesiadapid, *Stehlinella*. Of the tree-shrews, the Tupaiidæ possibly diverged at about the same time and have persisted with little change down to the present day. The Anagalidæ (represented by the fossil *Anagale*)—in view of the fact that in many features they showed a closer approximation to lemurs—probably left the main stem at a slightly later stage.

The Notharctinæ (which include *Notharctus, Pelycodus,* and perhaps *Aphanolemur*) represent a side-line which must have emerged relatively early, for the fossil remains of the early members of this sub-family indicate that they were small and primitive mammals. There can be no doubt that the Notharctinæ were quite aberrant lemurs. Their evolutionary differentiation evidently took place on the American continent, and, in the development of a most unusual molar pattern as well as in certain other points, it must be concluded that they departed definitely from the main line of descent of the modern Lemuriformes. The lower teeth of a genus *Protoadapis* from the basal Eocene (? Palæocene) of France are believed to represent a form closely related to *Pelycodus*. If this is so, it provides evidence that the early Notharctinæ were distributed over both the Old and New Worlds.

Pronycticebus is of special interest because of its very generalized structure (so far as it is known from the skull and dentition). It almost certainly lies very close indeed to the main stem of evolution of modern lemuriforms, not showing the moderate specializations peculiar to the other members of the Adapinæ. *Adapis* had probably reached a rather more advanced stage before it branched off. Since it presents a few slightly divergent modifications such as the withdrawal of the lachrymal foramen into the orbit, some degree of molarization of the last premolar and the rather early development of a strong hypocone in the molars, this genus can probably not be regarded as directly ancestral to modern lemurs. But in any case the Eocene ancestor of the lemuriforms must have resembled *Adapis* very nearly.

Up to this stage, it is almost certain that the evolutionary progress of the Lemuriformes had taken place somewhere in Eurasia, a few aberrant forms such as *Pelycodus* and *Notharctus* wandering off to the American continent. But it is probable that soon after the appearance of *Adapis* the lemuriform stock migrated to the Southern Hemisphere, eventually to gain relative isolation in Madagascar. In the seclusion of this island —and, it may be supposed, freed from the controlling influence of Natural Selection which had hitherto confined their evolutionary differentiation to rather limited directions—the Lemuriformes produced an astonishing variety of forms. It is in this " phylogenetic riot " that the lemurs betray their true relationship to the mammalian stock which also gave rise to the higher Primates. For some of the lemuroid genera, such as *Archæolemur*, which appeared in Madagascar during Pleistocene times were so remarkably pithecoid in appearance that some authorities have argued that they mark the actual transitional stage from lemur to monkey. It is certain that such is not the case, because true monkeys were already in existence in Europe long before, during the Oligocene and probably even the Eocene age. Moreover, these " pithecoid " lemurs of Madagascar exhibit specializations of the skull and teeth which could hardly have formed a stage in the evolution of the Anthropoidea. But the very fact that the Lemuroidea evinced the same tendencies in evolution is of the greatest importance in assessing their ultimate relationship to the Anthropoidea.

Having reviewed thus briefly the evolutionary radiations of the lemurs, we are now in a position to give a broad definition of the Lemuroidea in terms of their developmental trends, and to do this we may conveniently adopt a tabular form.

The Lemuroidea form a sub-order of the Primates which is distinguished from the other sub-orders by the following evolutionary tendencies :

(*a*) *In the skull*—reduction of the premaxilla (except in *Chiromys* and the Plesiadapidæ)—enclosure of the ectotympanic within the bulla (except in the Lorisiformes)—divergent modifications of the entocarotid artery which either undergoes atrophy (Lemuriformes) or takes an unusual course through the foramen lacerum medium (Lorisiformes)—participation of the palatine (Lemuriformes) or ethmoid (Lorisiformes) in the medial wall of the orbit.

(*b*) *In the dentition*—reduction of the upper incisors (except in *Chiromys*, Plesiadapidæ, *Archæolemur*) with procumbency of the lower incisors and associated modification of the lower canine—specialization of P_2—molarization of $P\frac{4}{4}$.

(*c*) *In the limbs*—progressive development of a " biramous " type of manus and pes—reduction of the second digit in the manus—compression of the os magnum—reduction of the lunate—displacement of the os centrale—elongation of the anterior tarsal segment—compression of the mesocuneiform —relative elongation of the fourth digits associated with an unusual disposition of the small muscles of the manus and pes—retention of a claw on the second pedal digit.

(*d*) *In the brain*—production of axial rather than limiting sulci on the dorso-lateral surface of the cerebrum (*Perodicticus* forms a conspicuous exception)—inversion of the lateral geniculate body (*see* p. 151).

(*e*) *In the special sense organs*—development of a nocturnal retina—retention of the primitive number of endo-turbinals in the nasal cavity—retention of the rhinarium, with a median sulcus in the upper lip—specialization of the sublingua— marked development of conical papillæ on the pharyngeal part of the tongue—elaboration of the external ear.

(*f*) *In the alimentary tract*—elaboration of the colon pattern to form an " ansa coli."

(*g*) *In the reproductive system*—elaboration of the penis in the development of " grappling spurs " and a baculum— perforation of the clitoris by the urethra (in the Lorisiformes) —retention of a primitive epithelio-chorial placenta.

The assumption was made above that the Lemuroidea form a sub-order of the Primates. There are, however, some anatomists who claim that they show such divergent specializations that they should be completely separated from the Order which contains *Tarsius* and the Anthropoidea. But it is evident from the foregoing chapters that, in spite of these divergencies, there are a number of distinctive features in which the living lemurs closely approximate to the higher Primates—such as the relative size of the brain, the formation of retrocalcarine and true Sylvian sulci, the elaboration of the visual centres, the reduction of the olfactory apparatus, the development of flat nails instead of claws, the degree of mobility of the pollex and hallux, the conformation of the ossicles of the middle ear, the appearance of the ethmoid in the medial wall of the orbit (in the Lorisiformes), certain features of the placentation, etc., which, in the aggregate, do emphasize the Primate status of these animals. Perhaps still more significant (as already pointed out) is the evidence of the fossil lemurs of Madagascar. Wood Jones has indeed remarked that " monkeys came near to being evolved " independently from these lemurs. The very fact that they did develop such remarkably pithecoid characters, even though superimposed on fundamentally divergent lemuroid specializations, is convincing evidence that the Lemuroidea and the Anthropoidea as a whole were derived from a common basal mammalian stock endowed with evolutionary potentialities which could ultimately give expression to the same traits. Wood Jones is probably correct in his inference that the ancestral stock from which the Lemuroidea and Anthropoidea diverged was so primitive and generalized that it would be impossible by a study of somatic structure to distinguish its members from those of the basal eutherian stock which also gave rise to other mammalian Orders. But their subsequent evolutionary history proves that they must have been differentiated from their contemporaries by the possession of inherent tendencies which were later to

become revealed in the development of many characters distinctive of the Primates among all other mammals.

The question as to whether the Lemuroidea could have given rise to the Anthropoidea after they had separated from the basal Primate stock requires attention here. We have already seen that the whole of the Lemuriformes must be excluded from such an ancestral relation because of the odd construction of the tympanic region of the skull, with which are associated an obliteration of the middle lacerated foramen and the atrophy of the main continuation of the entocarotid artery. The Notharctinæ may be *a fortiori* ignored in this connection because of the peculiar mode of development of the fourth cusp in the molars. The Lorisiformes are also specialized in the course taken by the entocarotid artery at its entry into the skull, a more primitive condition being preserved in the higher Primates. Apart from these considerations, the dominating evolutionary tendencies of the Lemuroidea enumerated above (the significance of which is tremendously enhanced by the fact that many of them have evidently developed independently in the two main groups of the Lemuroidea after they had separated in their evolutionary history as far back as Palæocene times) are almost all diametrically opposed to the corresponding tendencies shown in the Anthropoidea. Hence the inference follows that the Anthropoidea entered upon their own line of evolutionary development and separated off from the basal Primate stock before the Protolemuroids became differentiated in virtue of their special phylogenetic tendencies. Thus it is probably not legitimate to speak of a " lemuroid " phase in the developmental history of the Anthropoidea.

Tarsioidea

Apart from inferences which may be drawn from skeletal remains, we know nothing of the anatomy of the soft parts of any of the numerous tarsioid genera with the solitary exception of *Tarsius*—the only genus which persists to the present day.

The fact that in its structural features *Tarsius* displays quite a number of resemblances to the higher Primates has obtruded itself upon the notice of many observers since this small animal was first described by Burmeister in 1846. So much so, indeed, that some authorities have advocated the inclusion of *Tarsius* in the Anthropoidea, or at least in very close association with them. Thus Hubrecht, on the evidence of placentation, would have grouped it with the monkeys in his scheme of classification. Pocock—relying, it seems, mainly on the features of the rhinarium and external female genitalia—proposed to include *Tarsius* and the Anthropoidea together in one group, the Haplorhini, in contradistinction to the Strepsirhini which include the lemurs. But if attention is given to the totality of the anatomical characters of *Tarsius* (on which the assessment of affinities should depend), it may be demonstrated that the structural evidence upon which these authors placed reliance is to quite a considerable degree offset by that derived from other anatomical systems.

In the first place (as we have seen), it is evident that many of the so-called " pithecoid " traits of *Tarsius* are the result of fortuitous resemblances. For example, the reduction of the snout is to a great extent apparent only, its posterior part being overlapped and concealed by the expanded orbits ; the restriction of the nasal cavities is dependent upon the enlargement of the specialized orbits and has come about in a manner quite different from the analogous changes in the Anthropoidea ; the reduction of the olfactory parts of the brain is a secondary consequence of these peculiar modifications of the skull ; the pithecoid appearance of the brain as a whole is mainly the result of the compression of the brain-case due to the large orbits and to an expansion of the visual cortex (the latter, however, being based on the foundation of a lateral geniculate body which is *inverted* as in the lemurs and not everted as in the higher Primates) ; the lateral geniculate body, in its turn, is elaborated (so it seems) under the influence of the stimulus provided by a differentiated area in the middle of the retina, which is fundamentally different from the foveal differentiation in the Anthropoidea ; the displacement of the foramen magnum on to the basal aspect of the skull and the participation of the alisphenoid in the posterior wall of the orbit may possibly be entirely secondary to the enormous development of the eyes. Hence it may be argued that in respect of these features *Tarsius* merely apes the apes without really approximating to them.

But, apart from these examples of evolutionary convergence, *Tarsius* does show in a number of points suggestive evidence of real affinity with the Anthropoidea. These may be enumerated as follows : the absence of a true rhinarium ; the conformation of the male and female external genitalia ; the erect incisors and the simplicity of the premolars ; the hæmochorial placenta associated with the development of a connecting stalk ; the course of the entocarotid artery relative to the bulla ; the formation of a tubular auditory meatus ; the preservation of a primitive condition in the manus ; and the simple structure of the tongue. It is of course possible that some of these characters have developed in the tarsioid stock independently of the Anthropoidea (as, for instance, is probably the case with the tubular meatus), but, even so, they are indicative of a community of origin from an ancestral form with potentialities in these particular directions.

On the other hand, structural resemblances to the Lemuroidea (and corresponding contrasts with the Anthropoidea) are shown in the size of the brain relative to the body-weight, in features such as the inversion of the lateral geniculate body, the rod type of retina, the elaboration of the external ear, the construction of the hind-limb (including the digital formula of the pes), the bicornuate uterus, and in certain features of early development such as the mode of formation of the amnion.

A careful study of these data, with a full appreciation of

their relative value in the problems of taxonomy, and due consideration for the peculiar specializations distinctive of the tarsioid stock as a whole, leads to the conclusion that Gadow was perfectly right when in 1898 he proposed a central sub-order of the Primates to include *Tarsius* and to

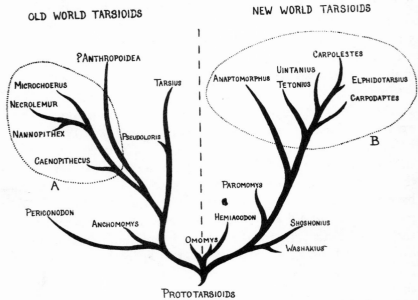

FIG. 88.—SCHEMA ILLUSTRATING THE EVOLUTIONARY RADIATIONS OF THE TARSIOIDEA IN THE OLD AND NEW WORLDS.

The main morphological trends of evolution are indicated, but not the geological horizons from which the remains of the extinct genera have been derived. A represents a group, the Microchœridæ, which is of particular interest because of the possibility that it may be closely related to the tarsioid stock which gave rise to the Anthropoidea. B represents a group of North American genera in which there is a prevailing tendency towards a high degree of dental specialization.

mark it off on the one side from the lemurs and on the other from the Anthropoidea. This classification has been strongly advocated by Elliot Smith and J. P. Hill, and is now generally accepted.

We may now consider what is known of the evolutionary radiations of the tarsioids. In Fig. 88 is a schema which is intended to direct attention to some of the main trends of structural specialization in the phylogenetic tree of the Tarsioidea, and again it should be noted that the geological horizons at which the various extinct genera have been found are not represented by the position they are given in this schema. Thus, for instance, *Omomys* and *Hemiacodon* are fossils which were discovered in middle Eocene deposits, while *Carpolestes* and *Elphidotarsius* are derived from upper Palæocene formations. But, judging by the degree of structural specialization which they show, it is probable that the former represent early offshoots from the main stem which survived to a later period with the preservation of primitive characters, while the latter represent the terminal stages of a definite but precocious phylogenetic trend associated with increasing specialization of the dentition.

It must be emphasized that some of these forms are known from very fragmentary remains, so that their allocation to the Tarsioidea may be uncertain. With this reservation, however, we may tentatively accept the diagnosis of the well qualified authorities who have made a study of them.

The place of origin of the Tarsioidea is obscure. Seeing that early forerunners of the various subdivisions of the Primates are so richly represented in Europe, while on the whole those found in America are rather specialized, it may be deemed probable that the Tarsioidea emerged from the basal Primate stock somewhere in the Eastern Hemisphere. On the other hand, fossil tarsioids (though almost all somewhat specialized) were in North America during the upper Palæocene period, and the earliest known European tarsioid (*Omomys*) is of lower Eocene age. Whatever may be the case, it is certain that quite early in the Tertiary epoch primitive tarsioids found their way both into Europe and into North America.* They may have undergone an independent

* The remains of an Eocene Primate (*Hoangbonius*) from China, discovered in 1916 by J. G. Andersson, have been referred provisionally by G. G. Simpson to the Anaptomorphidæ.

evolution in these two continents, or possibly successive waves migrated to both regions from some Asiatic source. Of the twenty or more genera of fossil tarsioids which have been described, rather more than half are of American origin. Only one genus is known which is common to North America and Europe, *Omomys*. That this form (of lower Eocene age) was very archaic is indicated by its retention of the generalized eutherian dental formula (with three incisors in one species) and the very primitive conformation of all the teeth. The remains of the tarsal elements of the Belgian species, however, show that even at this early stage of tarsioid evolution the hind-limb was well on its way toward the extreme type of specialization found in the modern tarsier.

On the American continent a few tarsioid genera besides *Omomys* have been discovered in which the dentition is quite generalized, but in a number of the New World fossils there is manifested a remarkable tendency towards a high specialization of the last premolar teeth, the molars on the contrary generally retaining the primitive trituberular form. This tendency reaches its culminating point in the genus *Carpolestes*, in which the last lower premolar is very large and develops a serrated shearing edge, but it also finds expression to various degrees in such forms as *Anaptomorphus*, *Tetonius*, *Uintanius*, etc. This prevailing phylogenetic trend in the American tarsioids has been emphasized in Fig. 88 by representing it as one of the end-products of their evolutionary radiation on this continent.

In Europe it is probable that the tarsioids developed along a number of different lines. Some, such as *Periconodon*, show rather unusual features in the cusp pattern of the molar teeth. Others showed a marked propensity for developing quadri-tubercular molars which in many respects approach the type found in higher Primates. The most important of these latter genera are *Microchoerus* and *Necrolemur*, which are sometimes included in a separate family, the Microchoeridæ. Apart from the dentition, *Necrolemur* shows in its skull certain pithecoid traits such as the position of the entocarotid foramen and the marked development of a tubular meatus. *Cænopithecus* may well belong to the same group, but the precise status of this fossil is a little uncertain. The fragmentary remains of its skull are sufficient to indicate that the orbits were not so large as in *Tarsius*, while the malar bone was stoutly built, the recession of the snout region very marked, and the symphysial region of the mandible was synostosed. On the whole, the skull of *Cænopithecus* seems to have been considerably more pithecoid in appearance than any other known tarsioid skull. *Pseudoloris*—in regard to its dentition—is among the most generalized of the European tarsioids. Like *Omomys*, it preserves three incisors. But the remains of the front of the skull show that in its orbital region it was practically as highly specialized as *Tarsius*. Indeed, it seems not unlikely that *Pseudoloris* represents the direct Eocene precursor of the modern *Tarsius*, and Teilhard de Chardin has suggested that it might well be renamed *Protarsius* (if the rules of nomenclature would permit it). It is not improbable, therefore, that *Tarsius* passed through the later stages of its phylogenetic history in the European area. In the line of *Pseudoloris* and *Tarsius*, the dentition has preserved many of its primitive eutherian characters unaltered to the present day.

Taking into account the palæontological evidence, and the anatomical data provided by *Tarsius*, we may define the Tarsioidea as follows :

The Tarsioidea form a sub-order of the Primates which is distinguished from the other sub-orders by the following evolutionary tendencies :

(a) *In the skull*—progressive enlargement of the orbits, accompanied by a compression of the nasal cavities*—participation of the ethmoid in the medial wall and the alisphenoid in the posterior wall of the orbit—displacement of the foramen

* This feature is indicated in all fossil tarsioids of which the skull is known at all, with the exception of *Cænopithecus*.

magnum on to the base of the skull—marked expansion of the bulla—the ectotympanic outside the bulla and produced into a tubular meatus—entrance of the entocarotid artery through the bulla wall—obliteration of the foramen lacerum medium.

(b) *In the dentition*—reduction of the incisors usually to two or one (the lower incisors disappear altogether in *Tetonius* and *Necrolemur*)—enlargement to varying degrees of the canines—retention of simple premolars and early loss of the first premolar—marked specialization of P$\frac{4}{4}$ (in American tarsioids)—molars either retaining the primitive tritubercular form (American tarsioids, *Pseudoloris*, *Tarsius*, etc.), or becoming quadritubercular and even multitubercular (Microchœridæ).

(c) *In the limbs*—the retention of a primitive structure in the fore-limb, including the digital formula and the disposition of the carpal elements—progressive specialization of the hind-limb* in relation to saltation, associated with fusion of the tibia and fibula, extreme lengthening of the anterior tarsal segment and relative elongation of the fourth pedal digit—retention of claws on the second and third pedal digits.

(d) *In the brain*—a broadening of the cerebrum as a whole with expansion and pronounced structural differentiation of the visual cortex and formation of a prominent occipital pole—inversion of the lateral geniculate body—retention of very primitive features in the cerebral commissures and the cerebellum.

(e) *In the special sense organs*—development of a nocturnal retina with a structural differentiation at its central point of an exceptional nature—reduction of the endo-turbinals following on the constriction of the nasal cavities by the large orbits—disappearance of a true rhinarium—retention of a primitive form of sublingua—elaboration of the external ear.

(f) *In the alimentary tract*—retention of a primitive colon pattern and mesentery.

(g) *In the reproductive system*—retention of primitive features in the external genitalia and in the bicornuate form of the uterus—acquisition of a hæmochorial placenta with the development of a connecting stalk and precocious vascularization of the chorion.

* This feature is present to some degree in all fossil tarsioids of which the limb skeleton is known.

That the earliest forerunners of the tarsioids—the Proto-tarsioids (as they may conveniently be called)—were very primitive and generalized mammals is sufficiently indicated by the anatomical features of their successors. The question arises whether in their phylogenetic emergence they separated from the basal Primate stock independently of the origin of the lemurs, or whether it can legitimately be said that they passed through a "lemuroid" phase before they became differentiated as a distinct group. In some respects—*e.g.*, in the characters of the hind-limb and the lateral geniculate body—they were evidently endowed with evolutionary potentialities identical with those of the Protolemuroid stock. But in so many points do they show divergent tendencies that it may be doubted whether the Tarsioidea and Lemuroidea are any more closely related than is implied by the statement that they were initially derived from a common basal Primate stock. Neither the Lemuriformes nor Lorisiformes could well provide an ancestral basis for the tarsioids, because of the specializations which they show in their cranial structure. Again, no fossil Primate is known in which the dentition is definitely "lemuroid" and at the same time sufficiently generalized to provide a point of departure for the evolutionary development of the tarsioid type of dentition. No doubt in respect of certain *separate* anatomical features, such as the nasal cavities and the placenta, a "lemuroid" phase must be postulated in tarsioid ancestry, in so far as these features have been developed by a modification of more primitive conditions which are still retained in the lemurs. But such a limited statement by no means implies that in the line of tarsioid descent there was a form which in the sum of its characters can be called a "lemur." On the contrary, a generalized common type which might have given rise to both the Tarsioidea and the Lemuroidea would presumably have been so primitive in respect of the brain, skull base, limb structure, etc., that it could hardly be distinguished from a generalized eutherian mammal. Hence it may be assumed that the Tarsioidea separated off from the basal Primate stock independently of the differentiation of the Protolemuroids with their own peculiar and distinctive tendencies for evolutionary development.

We may now consider whether the Tarsioidea could have provided a basis for the origin of the higher Primates, the Anthropoidea. We have seen that both these sub-orders show a number of common evolutionary tendencies. The American tarsioids can probably be excluded because of their general propensity for developing unusual dental specializations. On the other hand, among the European tarsioids are some—the Microchœridæ—which make quite a close approach to a pithecoid status in their dentition as well as in certain features of the skull. If the Anthropoidea represent the result of the progressive development of one branch of the tarsioid stock, therefore, it is not improbable that they came from the same stem which also gave rise to *Necrolemur*, *Microchœrus* and *Cænopithecus*. But this derivation must have occurred before the peculiar tarsioid specializations exemplified in the large orbits and modified hind-limbs had become manifested. This can only mean that the Anthropoidea branched off from the tarsioid stem close to its emergence from the Prototarsioid stock, for these sub-ordinal specializations are known to have put in an appearance in some of the earliest true tarsioids.

Anthropoidea

The Anthropoidea may be defined as a sub-order of the Primates which is distinguished from the other sub-orders by the following evolutionary tendencies :

(*a*) *In the skull*—progressive reduction of the snout region accompanied by a restriction of the nasal cavities—flexion of the basicranial axis and displacement of the facial skeleton below the front part of the neurocranium—great expansion of the neurocranium—complete rotation forwards of the orbital apertures—enlargement of the entocarotid artery—

participation of the ethmoid in the medial wall of the orbit—separation of the orbit from the temporal fossa by the expanded alisphenoid—ectotympanic forming a tubular auditory meatus and disappearance of the bulla as a prominent swelling (except in the Platyrrhines)—expansion of the frontal bones—displacement of the foramen magnum on to the base of the skull.

(*b*) *In the dentition*—incisors spatulate in form and reduced to $\frac{2}{2}$—canines often greatly enlarged, but may become secondarily reduced (as in Man)—premolars bicuspid and reduced to three or two—quadritubercular molars.

(*c*) *In the limbs*—retention of primitive features in both fore- and hind-limb, including the disposition of the carpal and tarsal elements and the digital formula—transformation of all the claws into flattened nails (except in the Hapalidæ and *Callimico*)—progressive elongation of the fore-limbs and retrogression of the pollex in the more arboreal forms.

(*d*) *In the brain*—great expansion of the cerebral hemispheres, which become richly fissured—characteristic development of the central sulcus and the lunate sulcus—very marked reduction of olfactory neural mechanisms and a corresponding differentiation of the visual apparatus—elaboration of the cerebellum—eversion of the lateral geniculate body (incomplete, however, in Man and the large anthropoid apes).

(*e*) *In the special sense organs*—differentiation of a macula and fovea at the central point of the retina—reduction of the external ear—retention of primitive features in the tongue—disappearance of the true rhinarium—marked reduction of the turbinate processes in the nasal cavity.

(*f*) *In the alimentary tract*—differentiation of the colon into ascending, transverse and descending portions—development of a vermiform appendix in the higher types (but absent in monkeys).

(*g*) *In the reproductive system*—retention of primitive features in the external genitalia, except for a small baculum in the monkeys—corpus uteri single—hæmochorial placenta with the development of a body-stalk—" spontaneous " development of the allantois and very precocious vascularization of the chorion.

Although palæontology has furnished a considerable amount of information regarding the later evolutionary radiations of the higher Primates, it has yielded surprisingly little evidence in regard to the actual origin of the pithecoid stock. As we have seen, some of the fossil tarsioids and lemuroids show in their structure a tentative approach to monkeys; but none of these, it appears, can represent the ancestral stock from which the latter were derived.

The study of the anatomy of living forms gives some indication of the primitive character of such an ancestral stock. In this regard, the vexed question of the Hapalidæ is of paramount importance. The question is vexed because persistent attempts have been made to represent this group as a very specialized and aberrant offshoot from the Cebidæ, the suggestion being that their primitive features are only apparent because they are the result of secondary retrogression. This refers especially to the sharp claws which characterize the manus and pes of these small monkeys.

We have argued in a previous chapter that the retention of claws in this case is truly primitive, and provides strong evidence for the supposition that the ancestral pithecoid type had probably not developed flattened nails except on the hallux. But it has been implied, for instance, that because the dental formula of the marmosets is the result of retrogressive changes, therefore it may be assumed that the presence of sharp claws is also a retrogressive trait. This implication does not take account of the fact that the claws are equally well developed in *Callimico*—one of the Cebidæ—in which the primitive Platyrrhine dental formula is retained. Apart from this, however, it has been pointed out in another connection that such a line of argument is highly fallacious. For nothing is more certain than the fact that a specialization in one part of the body is frequently associated with the preservation of archaic characters in other parts. Indeed, it may be said that the ability to preserve a primitive and generalized structure of the body as a whole is often directly conditioned by the adoption of a marked degree of specialization in one particular anatomical system.

But, in fact, the Hapalidæ exhibit so many obtrusively primitive structural characters that there can be little doubt that they do represent an offshoot from the very base of the Platyrrhine stem. We may recall (besides the presence of

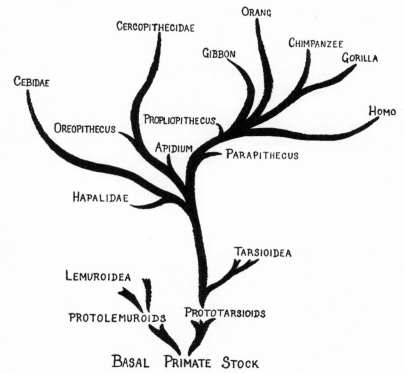

FIG. 89.—SCHEMA ILLUSTRATING THE EVOLUTIONARY RADIATIONS OF THE ANTHROPOIDEA.

The main morphological trends of evolution are indicated, but not the geological horizons from which the remains of the extinct genera have been derived.

sharp, compressed claws) the simple structure of the brain with its smooth neopallium and unelaborated cerebellum, the disposition and proportions of the tarsal elements, the musculature of the limbs, the incomplete excavation of the foveal spot in the retina, the conformation of the stapes, the simplicity of the intestinal tract, the remarkable development of complex scent glands in the perineal region, and the normal occurrence of plural gestation. Nevertheless, in many respects these small monkeys show in their general structure a great advance as compared with the lower Primates, especially in the relative development of the cerebral hemispheres.

In outlining the evolutionary radiations of the Anthropoidea, we may refer to the schema shown in Fig. 89, which is given to indicate pictorially the main conclusions to which our survey of the anatomical evidence has led us.

The mandible of a primitive representative of the Anthropoidea—*Parapithecus*—is known from lower Oligocene deposits of Egypt, and it is associated with the remains of a contemporary form, *Propliopithecus*, which is considerably more advanced and probably closely related to the immediate ancestors of the modern gibbon. These discoveries suggest that monkeys were already in existence at an even earlier date— that is, in Eocene times.

In the previous section we recognized the possibility that the Anthropoidea were derived from the earliest representatives of the Tarsioidea, possibly from the same group which also gave rise to the Microchœridæ. If this is so, the common ancestor of the Microchœridæ and the Anthropoidea must have been far more primitive than any known tarsioid. In the first place, the hind-limb in its structure presumably approximated quite closely to that of a generalized eutherian mammal. Again, the evidence of cerebral anatomy shows clearly that the brain of such a common precursor must have been relatively small and unelaborated. The divergent development of the geniculate bodies in *Tarsius* and the monkeys indicates that they could only have been derived from a structure which in its simplicity paralleled that of the tree-shrews. In this case, the pronounced development of the visual areas of the cortex, which is so characteristic of the living Anthropoidea, was not at that time manifested. Lastly, the evidence of the nasal cavities suggests that the Tarsioidea and Anthropoidea separated from a basal stock in which these cavities were fairly large, from which, again, it may be inferred that the olfactory parts of the brain were still quite well developed. It seems probable, from a consideration of data of this kind, that the common ancestral type may have possessed a skull which in its general proportions and conformation was not unlike that of the primitive lemuroid *Pronycticebus*.

Parapithecus lends some support to the conception of a tarsioid ancestry of the Anthropoidea, for there are certain features in the dentition and the conformation of the mandible of this fossil genus which recall characteristic tarsioid traits. But it does more than this. *Parapithecus* is claimed by some authorities to be an early representative of the Anthropomorpha, and thus a true anthropoid ape in contradistinction to a monkey. The possibility arises, therefore, that the anthropomorphs may have had a direct origin from a tarsioid ancestor without the intervention of stage which could be strictly termed Platyrrhine or Catarrhine. But the evidence is too slight to make more than a surmise, and the solution of such a question must be largely a matter of definition. *Parapithecus* may have the dentition of a primitive anthropoid ape, but if the rest of its anatomy were available for study it might be concluded on the basis of its skeletal, cerebral, placental and other features that it should be more correctly termed a Catarrhine monkey. There can be no doubt that in their phylogenetic history the anthropoid apes must have passed through stages in which, for instance, the brain, skull, placentation and tail were of the Catarrhine or Platyrrhine type, even though certain specializations which are distinctive of the existent monkeys (and obviously later acquired) would have been absent in such intermediate forms.

That the Platyrrhines left the main stem of evolution earlier than the Catarrhines is sufficiently indicated by their preservation of more primitive features in the dentition, tympanic region of the skull, placentation, etc., and by the generalized characters of the Hapalidæ. It may be supposed also that the Catarrhines passed through a Platyrrhine phase, but separated from the stock which gave rise to the modern New World monkeys at quite an early stage. This seems more probable than the alternative hypothesis that the Catarrhine and Platyrrhine monkeys evolved independently from separate tarsioid stocks. But we must certainly concede this as a possibility, simply because we are far too ignorant of the mechanism of evolution to put an arbitrary limit to the possibilities of parallelism. In this case, however, the detailed anatomical similarities between the two groups of monkeys are so numerous that we are entitled to postulate a common pithecoid ancestor in the absence of serious evidence to the contrary. Perhaps the only suggestive evidence for a diphyletic origin is that which depends on the geographical argument. This argument arises from the observation that the Catarrhines and Platyrrhines (recent and fossil) are sharply confined to the New and Old Worlds respectively, and that tarsioids were common to both Hemispheres. But the facility with which the Eocene fauna was evidently able to migrate between the Old and New Worlds, and the incompleteness of the palæontological evidence, render such a line of reasoning extremely insecure.

It is probable that the Platyrrhines separated from the stem which gave rise to the Catarrhine monkeys, the anthropoid apes and Man at a fairly early stage—when the brain was no more highly developed than it is in the marmosets, and when the digits (with the exception of the hallux) still bore sharp claws. It is also probable that the Catarrhine monkeys diverged, not very much later, from the line of evolution which led on to the anthropoid apes and Man, for, besides the evidence of the fossil *Parapithecus*, the Catarrhine monkeys have indulged uniformly in a number of rather distinctive specializations affecting the dentition, the development of cheek-pouches and callosities, and certain features of the brain such as the complete eversion of the lateral geniculate body, all of which suggest strongly that they have passed through quite a long history of evolutionary independence.

The earliest known fossil Platyrrhine monkey is represented by the genus *Homunculus*, which was found in Miocene deposits of Patagonia. This fossil gives no evidence bearing on Platyrrhine origins, for it represents a form which is already well differentiated and closely comparable with the smaller members of the existing Cebidæ.

Two early fossil Catarrhine genera of significance may be noted in connection with the phylogeny of the Old World monkeys. One of them—*Apidium*—is known by three molars and one premolar from the lower Oligocene of Egypt, and probably marks the beginning of the separation of the Cercopithecidæ from the generalized forerunners of the higher Primates. The other form is *Oreopithecus*—a lower Pliocene monkey—which can be fairly definitely relegated to the Cercopithecidæ.

From the data which have been thus briefly reviewed, it may be inferred that the Platyrrhines, Catarrhine monkeys and Anthropomorpha had become separated at least by the beginning of Oligocene times, and not improbably at an earlier stage during the Eocene period.*

The fossil remains of *Propliopithecus* and *Pliopithecus* from lower Oligocene and lower Miocene horizons respectively indicate that the anthropoid apes were certainly differentiated at this early stage, and there is some reason for believing that both these genera were closely related to the immediate ancestral stock of the gibbons.

With later stages in the evolution of the higher Primates we are not concerned in this thesis, for the problem of the phylogenetic differentiation and the interrelationships of

---

* Pilgrim, on the basis of some molar teeth which he discovered in upper Eocene deposits of Burma, has described a small anthropoid —*Pondaungia*. He believes this form to mark the commencement of the anthropomorph line of evolution and that it fills a gap between *Propliopithecus* and some Eocene tarsioid. But he points out that the teeth are badly weathered and his interpretations are quite tentative.

the anthropoid apes and Man has been reviewed in great detail by many competent authorities (*e.g.*, Elliot Smith, Gregory, Keith, etc.). We may note, however, that the emergence of the anthropoid apes and Man was marked by the adoption of an orthograde posture associated with such features as the loss of the tail, and by the progressive expansion of the higher functional levels of the brain—especially the neopallium ; but while the former have manifested an evolutionary tendency towards increasing specializations of the limbs in association with arboreal activities, the latter has preserved to a greater extent a primitive and generalized structure in this respect. We would refer to the evidence of the foot skeleton (*vide supra*, p. 131), which indicates fairly conclusively that the human stock separated from the anthropoid ape stock at a phase in which the body-weight of the common ancestor had not yet exceeded that of the modern gibbon, and that many of the remarkable similarities in structure between Man and the gorilla (for example) must therefore owe their origin to an evolutionary parallelism which dates back at least to this phase. The abundant remains of fossil anthropoid apes in Europe, Asia and Africa (*Dryopithecus, Sivapithecus, Palæopithecus, Australopithecus*, etc.) bear witness to the prolific nature of the evolutionary development of this Primate group all over the Old World during the later part of the Tertiary period. While most of these fossils have been regarded as bearing a not very distant relation to the recent anthropoid apes, some at least exhibit certain human features which make it probable that they are early derivatives of a stock which may also have given rise to Man's ancestors.

The later stages in Man's evolutionary history are represented by fossils such as *Pithecanthropus, Sinanthropus* and *Eoanthropus*, which have corroborated in a most remarkable way the inferences drawn by anatomists from the study of the structure of Man and the anthropoid apes. Such palæontological evidence leads inevitably to the conclusion that the progenitors of the Hominidæ, even if they avoided the specializations distinctive of the modern large apes such as the lengthening of the arms, the atrophy of the thumb and the hypertrophy of the canines, must have possessed features of the skull and jaws, teeth, brain, and limbs by which they would be quite consistently referred to the category of the anthropomorphous apes. That Man has been derived from a form which—without imposing any strain on commonly recognized definitions—can be properly called an "anthropoid ape" is a statement which no longer admits of doubt.

General Summary of the Evolution of the Primates

It is apparent that the phylogeny of the Primates has been no simple matter of four or five straight lines of evolutionary development beginning in an ancestral form of lemuroid type and culminating at successive levels in the modern lemurs, *Tarsius*, Platyrrhine and Catarrhine monkeys, anthropoid apes and Man. On the contrary, the living members of this Order must be recognized as the very few survivals of a whole series of evolutionary radiations which diverged *ab initio* from an extremely generalized stock of placental mammals and which incorporated quite wide potentalities for specialization in diverse directions. It is not impossible that from the undifferentiated basal eutherian stock a great number of quite separate groups emanated in the earliest phases of its evolution, bearing no closer relation to each other than is implied in their common origin from such a basal stock and in the fact that they manifested potentialities for development along the general lines which are taken as distinctive and diagnostic of the Primates among other mammalian Orders. Thus many of the various subdivisions of the recent and fossil Primates which have been recognized as different families or genera may conceivably be polyphyletic in their ultimate origin, and it is possible that a more extended palæontological record may necessitate the creation of more than three sub-orders to allow of a natural classification of all the genera which are known. Nevertheless, on the evidence at present available, the known genera of the Primates do allow themselves to be grouped in three main subdivisions or sub-orders, each of which is characterized by the manifestation of certain definite and fairly limited tendencies in evolutionary development. The observed facts may therefore be best expressed by regarding them as the derivatives of three main lines of evolutionary radiation which, as it were, crystallized out from the genetically homogeneous and undifferentiated matrix of a basal Primate group.

We may, then, conceive the general evolutionary history of the Primates to be somewhat as follows : At the end of Mesozoic times—either at the end of the Jurassic or the beginning of the Cretaceous period—a basal stock of generalized placental mammals emerged (probably from still earlier primitive mammals of the type recognized in the extinct Pantotheria) and formed the foundation for the subsequent differentiation of all the various Orders of eutherian mammals which are to-day known. By the beginning of the Tertiary period these Orders were already distinct, and their earliest representatives were probably very similar in type to the remarkably interesting Cretaceous placental mammals of Mongolia which have been described by W. K. Gregory and G. G. Simpson—*Zalambdalestes* and *Deltatheridium*.

In this initial " splitting up " of the basal placental stock, a group of generalized mammals appeared which were endowed with or acquired potentialities for developing in the particular direction which led ultimately to the appearance of the Primates. This group may be termed the Basal Primate Stock, and it comprised primitive types which at first could not have been distinguished from the basal stocks of other Orders except by reference to their evolutionary tendencies (which became successively manifested during the Tertiary epoch). These tendencies, indeed, provide a convenient basis for a definition of the Order. Thus it may be said that the Primates form an Order of the class Mammalia, which is distinguished from the other Orders by the following evolutionary tendencies :

(*a*) *In the skull*—progressive reduction of the facial part of the skull, with recession of the snout region and restriction of the nasal cavities—expansion of the neurocranium—forward rotation of the orbital apertures—completion of a postorbital bar—participation of the ethmoid in the medial wall of the orbit (except in the Lemuriformes)—displacement of the foramen magnum towards the base of the skull—formation of an osseous floor of the tympanic cavity from a process of the petrosal bone.

(*b*) *In the dentition*—reduction of the incisors usually to $\frac{2}{2}$—the premolars reduced to $\frac{3}{3}$ or $\frac{2}{2}$, and preserving on the whole a simple character—molars gradually assuming a quadritubercular pattern by the upgrowth of a true or pseudohypocone.

(*c*) *In the limbs*—preservation of a generalized structure in the fore- and hind-limb, including retention of the clavicle, a generalized arrangement of the carpal and tarsal elements (except in the lemurs), and pentadactyly with a primitive digital formula—radio-ulnar joint capable of free rotation—fibula usually remaining free—transformation to a varying degree of claws into flattened nails—enhancement of the grasping power of the pollex and hallux, with free mobility of these digits—increasing differentiation of the small muscles of the manus and pes.

(*d*) *In the brain*—progressive expansion of the fore-brain and especially of the neopallium—marked degree of fissuration of the neopallial cortex with the appearance of a retrocalcarine sulcus and a true Sylvian fissure—backward projection of the occipital lobe of the hemisphere—reduction of olfactory neural mechanisms and a corresponding elaboration of the visual apparatus—production of a six-layered lateral geniculate body of which the superficial layers are characteristically large-celled.

(*e*) *In the special sense organs*—development of a retina either of the nocturnal or diurnal type and, in the latter case, the differentiation of a fovea—incomplete decussation of the optic tracts—reduction of the turbinate processes in the nasal cavity—retention of generalized features in the tongue (except in the Lemuroidea)—reduction of the external ear (except in lemurs and *Tarsius*).

(*f*) *In the alimentary tract*—retention of generalized mammalian characters with a simple colon pattern (except in the Lemuroidea).

(*g*) *In the reproductive system*—external genitalia devoid of elaboration (except in the Lemuroidea)—uterus either bicornuate or single—penis pendulous—permanent descent of the testes into the scrotal sac—progressive acceleration of the early stages of embryonic development leading to a rapid vascularization of the chorion and the early establishment of the placental circulation—placenta of the hæmochorial type (except in the lemurs).

From the basal Primate stock charged with these general developmental predispositions, it is probable that the precocious dissociation of a separate and isolated group marked the commencement of the phylogenetic history of the lemurs. This group—which is here designated the Protolemuroid stock—diverged rather sharply from the other derivatives of the basal Primate stock and showed a marked proclivity for a series of specializations of an exceptional type affecting many parts of the body. From the Protolemuroids emerged independently the Lorisiformes and the Lemuriformes. From the latter, again, the line represented by the Plesiadapidæ and *Chiromys* branched off at a very early stage. The initial phases in the evolutionary differentiation of the Lemuriformes are illustrated to-day by the Tupaioidea, an early offshoot from the main stem which has preserved the primitive characters of the group to an extraordinary degree. It is probable that these initial stages took place somewhere in the Eurasian land-mass at least as early as Palæocene times, for, before the end of this geological period had been reached, aberrant lemuriforms represented by the early Notharctinæ —*Pelycodus*—had migrated to America, and from this stock the genus *Notharctus* almost certainly arose. The less specialized trends of lemuriform evolution are exemplified by the European fossils *Pronycticebus* and *Adapis*, and of the latter genus some species have been found in considerable numbers in Eocene deposits. After Eocene times the geological history of the Lemuriformes is a complete blank up to the Pleistocene. They are then found in Madagascar, to which they are confined at the present day. There is no palæontological record of the Lorisiformes at present available, in spite of the fact that comparative anatomy indicates that the living lorisiforms must be the end-products of a long history of evolutionary independence dating from the time when the Protolemuroids first commenced their differentiation. This gap in the geological record is the more astonishing in view of the fact that the Lorisiformes are to-day distributed over a very wide area in the Old World.

Independently of the origin of the Protolemuroids and avoiding the deviating tendencies for specialization displayed by them, there arose from the basal Primate stock a group of primitive mammals which showed a marked conservatism in the retention of a more generalized bodily structure, and at the same time incorporated potentialities for a progressive development of the brain, skull, teeth, limbs, rhinarium, etc., of the type which became finally manifested in the Tarsioidea. These may be termed the Prototarsioids, and they must have come into being by the very beginning of Palæocene times, if not during the Cretaceous period. Their differentiation led to the appearance of a number of separate genera which were distributed over both Old and New Worlds by the end of the Palæocene. The dominating tendencies in the evolution of the tarsioids were the rapid enlargement of the eyes, and the aberrant modification of the hind-limbs in adaptation for leaping among the branches. These specializations developed extremely early in the phylogenetic history of the group. But it may be presumed that the earliest derivatives of the Prototarsioids did not as yet display these features, and at the same time they may have been sufficiently advanced structurally to deserve recognition as true tarsioids in virtue of an incipient expansion of the brain, the shutting off of the orbit from the temporal fossa, the acquisition of a mobile upper lip, the structure of the tympanic region of the skull and so forth.   Of these early true tarsioids, one branch (which also gave rise ultimately to the Microchœridæ) may have provided a foundation for the emergence of the Anthropoidea.   It is certain that no closer relationship between the Anthropoidea and the Tarsioidea can be claimed than is implied in such a common origin from the very base of the tarsioid stem.   Quite possibly, indeed, a more complete palæontological record will finally demonstrate that these two sub-orders are separate and distinct radiations from the basal Primate stock and thus have followed completely independent lines of development—a conclusion which would harmonize satisfactorily with the strongly contrasting trends of evolution which are shown in each group.   But the evidence at present available at least does not conflict with the conception of a common origin from an ancestral form which could legitimately be called " tarsioid."

The line of evolution of the Anthropoidea has been marked by the successive branching off of specialized groups from a central stem in which a progressive expansion of the brain has been accompanied by the retention of a bodily structure of a remarkably generalized type.   It is this main stem which culminated in the appearance of Man himself.

One of the most impressive features which has emerged from a detailed discussion of the anatomical evidence is the demonstration of the part which has been played by parallel evolution in the phylogeny of the Primates.   It is clear that the three main subdivisions of the Order followed out from the very beginning of Tertiary times a similar evolutionary course in respect of more than one anatomical system.   Within the limits of the sub-orders, also, there are contrasting lines of collateral descent—such as the Lemuriformes and Lorisiformes, the Catarrhine and Platyrrhine monkeys, the large anthropoid apes and Man—which provide examples of the closest parallelism, the result evidently of the manifestation of identical developmental trends.   Moreover, it has been made evident that this parallelism may involve very small and apparently trivial details. It may lead to the development of specializations which are not always definitely attributable to environmental needs—that is to say, there may be no positive evidence that the survival of the species actually demands them.

General biologists have hitherto hardly given adequate attention to the nature of the evidence which the science of comparative anatomy has to offer in the solution of evolutionary problems. This evidence leads inevitably to the conclusion that evolutionary parallelism has been a much more common phenomenon than is usually recognized, and it must obviously be taken into account in any philosophical study of evolutionary principles.

Clearly, the parallelism which has been noted in the Primates is merely an expression of the thesis that the descendants of a common ancestor always tend to evolve along similar lines. This was first explicitly stated by Osborn in his observation that " the *same* results appear independently in descendants of the *same* ancestors,"* and it has since been frequently quoted by those who have attempted to reconstruct phylogenetic histories from purely anatomical data. It is perhaps not fully realized that Osborn's dictum is but another way of expressing the principle of Orthogenesis. This principle embodies the conception that evolution is the manifestation of an inherent tendency in the germ-plasm to vary along-definite and limited lines ; the modification of an organism is not due to the natural selection of apparently fortuitous variations which may occur in any direction, but rather to a process of continuous change which is taking place in the germ-plasm itself. From this point of view, the evolutionary development of new somatic characters may occur in the face of, or at least independently of, direct environmental influence. No doubt, of course, the powers of Natural Selection may lead to the early extinction or abortion of a stock which

* H. F. Osborn : Science, vol. xxvii., 1908.

manifests a tendency to the development of harmful characters, or favour the survival of a stock in which there is an inherent tendency to evolve in an advantageous direction. Palæontologists have, however, brought forward evidence which suggests that the progressive evolution of certain characters may persist over prolonged periods although habits and habitat change considerably, and in spite of the fact that they do not harmonize with the environment. This process may ultimately lead to the complete extinction of a widely distributed and large group of animals.

It seems certain that the instances of parallelism in the evolution of the Primates which have been brought to light in the preceding chapters are to be interpreted satisfactorily only by the conception of definite predetermined trends of development—that is, by the conception of Orthogenesis. This conception puts the onus of evolutionary progress more on the germ-plasm and regards the influence of the environment as of somewhat secondary importance. Hence it seems to intensify the mysteries of the germ-plasm, which (it implies) is endowed from the beginning with countless potentialities for evolution in definite directions. It becomes, therefore, increasingly difficult to conceive of evolution as being fundamentally merely a matter of action and reaction between the physico-chemical factors of the environment and those of a passive or at least a neutral and completely plastic organism. For this reason, Orthogenesis is apt to be dismissed rather abruptly as a " vitalistic " principle complicating in an unwelcome manner the mental pictures which biologists have striven to elaborate under the influence of mechanistic ideas. But if the mysteries of the living and evolving germ-plasm are even deeper and more enigmatical than we have been inclined to believe, it were better to recognize the fact.

INDEX

Abel, O., 75, 84, 91, 95, 120, 234
Adapidæ, incisors, 79
    lachrymal bone, 49
    premaxilla, 49
    skull, 48, 255
Adapinæ, 37
    foramen magnum, 51
    orbit in, 185
    symphysis menti in, 48
" Affenspalte," 11
Alimentary tract, Anthropoidea, 273
    Lemuroidea, 261
    Primates, 284
    Tarsioidea, 270
Anagalidæ, dentition, 224
Anaptomorphidæ, dentition, 34
Ancestry, lemuroid, 17
    pithecoid, 17
    tarsioid, 17
Ant-eater, great, 146
Anthropoid apes, cæcum, 205
    definition, 14
    liver, 206
    reproductive system, 210
    thumb, 138
Anthropoidea, 22
    alimentary tract, 273
    ancestral stock, 262
    brain, 163, 273
    carpus, 125
    dentition, 93, 273
    lachrymal bone, 63
    limbs, 120, 273
    mandible, 64
    nasal cavities, 179
    os centrale, 125
    penis, 210
    placenta, 218
    reproductive system, 210, 273
    retina, 183, 186

Anthropoidea, skull, 61, 272
    special sense organs, 273
    thumb, 126
    tongue, 195
Anthropomorpha, 25
    definition, 15
Appendix, *Cercopithecus*, 205
    Chiromys, 204
    Galago, 204
    Lemur, 204
    Perodicticus, 204
    Phascolomys, 205
Arboreal man, 103
Archæolemuridæ, symphysis menti in, 48
Arms of human embryo, 3
Arteria promontorii, in Lemuriformes and Lorisiformes, 47
    in Primates, 54
    in Tarsioidea, 57
Artery, entocarotid, in Lemuriformes and Lorisiformes, 47, 55, 60
    in Tarsioidea, 57
    in tree-shrews, 226
    stapedial, 47, 66
    in tree-shrews, 226
Assessing affinities, 8
Aye-aye, *see Chiromys*

Baboons, 29, 30
Baculum, in *Cebus*, 210
Banks, E., 228
Basi-cranial axis, 41
Beattie, J., 129, 123, 207, 212
Bilophodont molars in Catarrhines, 16
Bolk, 168
Boule, Prof., 50
Boyd, J. D., 175

" Brachiation," 25, 121
Brain, Anthropoidea, 13, 163, 273
    Catarrhine monkeys, 166
    Cebidæ, 165
    Centetes, 142
    Chiromys, 100, 109, 117, 132,
        135, 155, 162
    evidence of the, 141
    Lemuroidea, 145, 261
    marmoset, 13
    Primates, 283
    simian sulcus, 11
    Tarsioidea, 156, 158, 165, 270
    tree-shrews, 234
    Tupaioidea, 234
Brodmann, K., 155
Burmeister, H., 264

Cæcum in Anthropoidea, 205
    in Lemuroidea, 203
    in monkeys, 204
    in *Tarsius*, 204
Callosities, ischial, 16, 17, 27, 213
Canine teeth, 70
    Anthropoidea, 93
    man, 11
    Tarsioidea, 88
Capuchin monkeys, 31
Carlsson, A., 246
Carpus of Anthropoidea, 125
    of Lemuroidea, 110
    of Tarsioidea, 116
    of Tupaiidæ, 227
Carter, J. Thornton, 100
Catarrhine monkeys, 16, 27
    and Platyrrhine, distinc-
        tion, 177
    brain, 166
    cæcum, 205
    definition, 16
    dentition, 93
    fossil, 29
    ischial callosities, 16, 27,
        213
    limbs, 122, 127
    nostrils, 176
    ossicles, 192
    placentation, 218
    Platyrrhine phase, 278
    progenitors, 16

Catarrhine monkeys, reproductive
        system, 211
    retina, 187
    tongue, 195
Cebidæ, 30
    brain, 165
    claws or nails, 123, 274
    reproductive system, 209
Cercopithecidæ, 28
    premolars in, 94
Cercopithecinæ, 29
Chardin, Teilhard de, 84, 119, 234
Cheek pouches in Catarrhines, 27
Cheirogaleinæ, 36
Chimpanzee, 25
Chiromyidæ, 36
Chiromyoidea, 36, 256
Chiromys, claws or nails, 35, 110,
        132, 135
    dentition, 84
    limbs, 109
    reproductive system, 211
Cingulum, 71
Claws and nails, distinction, 230
Colic loop (Lemuroidea), 201
Colon in *Chiromys*, 201
    in *Chirogaleus*, 201
    in *Chrysothrix*, 201
    in *Galago*, 201
    in *Hapale*, 202
    in *Indris*, 201
    in *Loris*, 201
    in *Microcebus*, 201
    in *Nycticebus*, 201
    in *Perodicticus*, 201
    in *Tarsius*, 201
    in *Tupaia*, 201
Cooper, Forster, 191
Cope, E. D., 72, 162
Cynomorpha, 27

Darwin, 1, 139
Daubentoniidæ, 36
Definitions, 14
Deltatheridium, 282
Dental formula, 70
Denticulated sublingua in Lemur-
        oidea, 194
Dentition, 8, 10, 70
    Adapidæ, 79

Dentition, Anagalidæ, 224
  Anaptomorphidæ, 34
  Anthropoidea, 93, 273
  Catarrhine monkeys, 93
  Cercopithecidæ, 94
  Chiromyidæ, 36
  Chiromys, 84
  Hapalidæ, 93
  Lemuroidea, 77, 261
  Man, 11
  Microchœridæ, 268, 272
  Pantotheria, 70
  Platyrrhine monkeys, 93
  Primates, 73, 283
    microscopical structure, 100
  Tarsioidea, 87, 270
  tree-shrews, 231
  Tupaiidæ, 231
Deuterocone, 75
Digestive system, evidence of the, 197
  in *Chrysothrix*, 199
  in *Colobus*, 193
  in *Semnopithecus*, 198
  in *Tarsius*, 199
  in Tupaiidæ, 239
Doran, A. H. G., 191
Dubois, E., 24
Duckworth, W. L. H., 206

Ear, external, in tree-shrews, 254
  in *Alouatta*, 191
  in Anthropoidea, 191
  in *Cebus*, 191
  in *Cheirogaleus*, 190
  in *Chiromys*, 190
  in *Galago*, 190
  in Hapalidæ, 191
  in *Lemur*, 190
  in Lemuroidea, 189
  in *Microcebus*, 189
  in *Nycticebus*, 190
  in Old World monkeys, 191
  in *Perodicticus*, 190
  in Platyrrhines, 191
  in *Tarsius*, 190
Échelle des êtres, 9
Ectotympanic bone, 47, 54
Edinger, Dr. Tilly, 152

Embryological evidence, 3
Entoconid, 73
Eocene period, 21
Eutheria, 20
Eutherian stock, 21
  generalized dentition, 70
Evidence of the brain, 141
  of the digestive system, 196
  of the limbs, 103
  of the reproductive system, 209
  of the skull, 40
  of the special senses, 173
  of the teeth, 69
Evolution of the Primates, general summary, 281
Evolutionary origin of the Primate limbs, 132
  parallelism, 6
  radiations of the Anthropoidea, 273
    of the Lemuroidea, 254, 261
    of the Primates, 253
    of the Tarsioidea, 263, 269
Eye, the, 182
  Adapis, 185
  Anthropoidea, 185, 273
  Cercocebus, 183
  Chimpanzee, 183
  Hapale, 184
  Lemuroidea, 185, 188, 261
  marmoset, 186
  Megacheiroptera, 185
  Microcebus, 188
  Notharctus, 185
  Nyctipithecus, 183
  Primates, 187, 283
  Ptilocercus, 244
  Tarsius, 184, 270
  Tupaia, 244

Falcula (Primates), 104
Fischers, E., 180
Fœtal membranes, 214
Foot of human embryo, 3
  of Anthropoidea, 273
  of *Dendrolagus*, 135
  of kangaroo, 135
  of Lemuroidea, 261

Foot of Tarsioidea, 270
Franz, 184
Frazer, J. E., 175

Gadow, 266
Galagidæ, 36
Genetic affinity, degrees of, 4
Genitalia, Anthropoidea, 210
 Lemuroidea, 210, 212, 261
 Tarsius, 210, 270
 tree-shrews, 243
Gibbon, 25
Gorilla, 25
Gorilloid ancestor, 3
 heritage, 138
Grandidier, G., 50
" Grappling spurs," 210
Gregory, W. K., 37, 49, 54, 77, 86,
 93, 98, 111, 120, 138, 280, 282
Grosser, O., 215
Guereza monkeys, 29

" Halbaffen," 34
Hapalidæ, 30
 brain, 163
 clawed digits, 123, 129
 colon, 202
 dentition, 93
 hallux, 129
 limbs, 121
 ossicles, 192
 primitive structures of, 275
 reproductive system, 212
 retina, 184
 sense of smell, 179
 tactile pads, 137
Haplorhini, 222
Henckel, K. O., 180
Hill, J. P., 123, 215
Hominidæ, 23
 derivation, 15
Homo neanderthalensis, 23
 rhodesiensis, 23
 sapiens, 23
Homunculus, 32
Howler monkey, 31
Hubrecht, A. A. W., 32, 264
Huxley, T. H., 139
Hylobatidæ, 25
Hypocone, 74

Hypoconid, 73
Hypoconulid, 73

Incisors, 70
Indrisidæ, 36, 112
Inguinal canals, 209
Irreversibility of evolution, 12
Ischial callosities, 16, 213

Johnston, T. B., 204
Jones, F. Wood, 28, 32, 103, 117,
 134, 138, 194, 205, 213, 262

Kangaroo, foot of, 135
Kaudern, W., 242
Keith, A., 120, 280
Klaatsch, H., 200, 202
Kollmann, M., 45, 178, 180
Kolmer, 185
Kornyey, 150

Langurs, 29
Lemoine, V., 84
Lemuridæ, 36
Lemurinæ, 36
Lemuroid ancestry, 17
Lemuriformes, and Tupaioidea, 249
 arteria promontorii, 47
 dentition, 99
 ectotympanic bone, 54
 geological history, 285
 nasal fossæ, 178
 snout of, 42
 tongue, 192, 194
 tympanic bulla, 46
Lemuroidea, 22, 34, 42, 254
 ancestral stock, 262
 brain, 145
 cæcum, 203
 colic loop, 201
 dentition, 77
 digestive system, 201
 ear, 189
 earliest progenitors, 258
 limbs, 106, 111
 liver, 206
 ossicles, 191
 reproductive system, 210
 skull, 42, 261
 special sense organs, 261

Lemuroidea, sublingua, 193
   tongue, 192
Lemurs, placentation of, 12
   nasal cavity, 178
Lesser tree-shrew, 147
Limbs, Anthropoidea, 120, 273
   Catarrhine monkeys, 127
   Cebidæ, 123
   evidence of the, 103
   evolutionary origin of Primate,
     132
   Hapalidæ, 121
   Lemuroidea, 106, 261
   Platyrrhine monkeys, 126
   Primates, 104, 113, 283
   Tarsioidea, 115, 270
   Tarsius, 105
   tree-shrews, 227
Lipotyphla, 222
Liver in Lemuroidea, 206
   in marmoset, 207
   in *Tarsius*, 207
Lorisidæ, 36
Lorisiformes, dentition, 99
   distribution, 285
   nasal fossæ, 178
   tongue, 192, 194
   tympanic ring, 46
Lorenz, G. F., 229
Lytta, in tree-shrews, 245

Macaque, pig-tailed, 28
Macroscelidoidea, 222
Macula, 183
Man, ancestry, 17
Mandible, Anthropoidea, 64
   Lemuroidea, 48
   Tarsioidea, 57
Mangabey monkey, 28
Manus, marmoset, 136
Marmoset, brain, 13
   scent glands, 179
   sublingua, 195
   tongue, 195
Marsupialia, 20
Matthew, W. D., 20, 33, 38, 84, 234, 250
Megacheiroptera, retina of, 185
Menotyphla, 222, 250
Mesoduodenum, 200

Mesozoic epoch, mammals in, 20
Metacone, 72
Metaloph, 76
" Metatarsi-fulcrumating," 129
Metatheria, 20
Microchœridæ, 34, 286
   dentition, 268, 272
Microscopical structure of Primate
   teeth, 100
Miocene period, 21
Missing links discarded, 13
Mitchell, P. Chalmers, 197, 248
Molar teeth, Anthropoidea, 93
   cusps of, 5
   Lemuroidea, 77
   Tarsioidea, 89
Morton, D. J., 108, 113, 120, 131, 138
Mouse-lemur, 145
Multitubercular theory, 10
Multituberculata, 19, 72, 74
Muscular system of the Lemuroidea, 246
   of the Tupaiidæ, 246

Nasal cavity, 177
   Chiromys, 178
   Hapale, 177
   Lemur, 177
   Tarsius, 177, 180
   tree-shrews, 244
Nasalis, 29
Nayak, U. V., 190, 207, 209
Negative reversal, 13
Neumayer, L., 162
New World monkeys, *see* Platyr-
   rhine monkeys
Nostrils, *Ateles*, 177
   Catarrhine monkeys, 177
   Cebus, 176
   Cercopithecus, 176
   Colobus, 177
   Platyrrhine monkeys, 177
Notharctinæ, 263
   aberrant lemurs, 259
Notharctus, lachrymal bone in, 49

Old World monkeys, *see* Catarrhine
   monkeys
Olfactory mechanisms, 143

Olfactory mechanisms, Anthropoidea, 176
    Catarrhine monkeys, 177
    Lemuroidea, 174
    Platyrrhine monkeys, 177
    scrolls, 177
        Orycteropus, 178
Oligocene period, 21
Ontogeny and Phylogeny, 3
Opposition of the thumb, 228
Orang-utan, 25
Orbit in *Adapis*, 185
    in *Notharctus*, 185
Orbito-temporal region, 56, 185
Oreopithecidæ, 30
Orimentary hypocone, 75
Orthogenesis, 237
Os centrale, 125
Osborn, H. F., 7, 72, 84, 138, 287
Ossicles of middle ear, Catarrhines, 192
        Hapalidæ, 191
        Lemuroidea, 191
        Tarsius, 192
        Tupaia, 245

Palæocene period, 21
Palæontological records, 4
Pantotheria, 19, 70
Papin, L., 178
Paracone, 71
Parallelism in evolutionary development, 6
Parapithecidæ, 26
Pelvic girdle, 111
Pen-tailed tree-shrew, 223
Penis in Anthropoidea, 210
    in Lemuroidea, 210
    in *Tarsius*, 210
Pentadactyly, 9
Philtrum, 175
Phylogenetic principles, 3
Pilgrim, G. E., 279
Pithecoidea, 27
Placenta, structure of, 213
Placentalia, 20
Placentation, Anthropoidea, 218, 273
    Catarrhines, 218
    Lemuroidea, 215, 261

Placentation, Platyrrhines, 218
    Tarsius, 217, 270
    Tupaia, 244
Platyrrhine monkeys, 27, 30
    brain, 166
    cæcum, 204
    dentition, 93
    ear, 191
        ossicles, 192
    limbs, 126
    nostrils, 177
    placentation, 218
    reproductive system, 210
    retina, 187
    skull, 61
    sublingua, 195
    tympanic bulla, 64
Pleistocene period, 21
Plesiadapidæ, age of, 38
    humerus, 107, 133
    incisors, 84
    skull, 51
Pliocene period, 21
Pocock, R. I., 31, 32, 190, 194, 209, 212, 264
Pondaungia, 279
Primates, alimentary tract, 284
    arteria promontorii, 54
    brain, 283
    definition, 22, 283
    dentition, 73, 283
    distribution, 19
    ꞉ ear, 189
    evolution, 281
    eye, 182, 283
    foot mechanism, 113
    limbs, 104, 113, 132, 283
    reproductive system, 209
    skull, 40
    stereoscopic vision, 187
    teeth, microscopical structure of, 100
Primordium maculæ, 184
Proboscis monkey, 29
Prosimiæ, 22
Prostate gland, 212
    in *Ptilocercus*, 242
    in *Tupaia*, 242
Protoconid, 71
Proto-lemuroid stock, 258, 285

Protoloph, 76
Proto-tarsioids, 271
Prototheria, 20
Pseudohypocone, 75
Ptilocercinæ, 223

Quaternary epoch, 21

Recent period, 21
Regan, C. Tate, 12, 101
Remane, A., 95
Reproductive system, anthropoid
    apes, 210
    Anthropoidea, 218, 273
    evidence of the, 209
    Hapale, 212
    Lemuroidea, 212, 261
    Primates, 209, 284
    Tarsioidea, 210, 270
    tree-shrews, 241
Retina in Catarrhine monkeys, 187
  in *Hapale*, 184
  in Lemuroidea, 185
  in Megacheiroptera, 185
  in *Nyctipithecus*, 184
  in Platyrrhine monkeys, 187
  in *Ptilocercus*, 244
  in *Tarsius*, 184
  in *Tupaia*, 244
  in vertebrate type, 182
Ruge, 206

Saki monkey, 31
Schlosser, M., 84
Schwalbe, G., 30
Scotoptic vision, 182
Scrotum in Primates, 209
Semnopithecinæ, 29
Simian sulcus, 11
Simiidæ, 25
Simpson, G. G., 19, 20, 70, 87, 233,
    234, 282
Skull, Adapidæ, 48, 255
    Anagale, 225
    Anthropoidea, 61, 65, 272
    evidence of, 40
    evolution of, 8
    Lemuroidea, 42, 55, 261
    Primates, 283
    Ptilocercus, 225

Skull, Tarsioidea, 56, 269
    tree-shrews, 225
Smith, G. Elliot, 11, 40, 105, 143,
    146, 155, 163, 173, 187, 266, 280
Sonntag, C. F., 192, 212
Special senses, evidence of the, 173
    sense organs, Anthropoidea,
    275
    Lemuroidea, 261
    Primates, 283
    Ptilocercus, 244
    tree-shrews, 244
    Tupaia, 244
Spider monkey, 31
Squirrel monkey, 31
Stehlin, H. G., 50, 55, 59, 79, 84,
    85, 234
Stereoscopic vision, Primates, 187
Strahl, 244
Strauss, W. L., 118
Strepsirhini, 222
Sublingua, in Lemuroidea, 193
  in marmoset, 195
  in *Tarsius*, 195
Sylvian fissure, 146
Symmetrodonta, 19
Symphysis menti in Archæole-
    muridæ, 48

Tænia coli, 204
Talonid, 73
"Tarsi-fulcrumation," 22, 32, 114,
    115, 119, 129
Tarsioidea, brain, 156
    definition, 269
    dentition, 87, 270
    limbs, 115
    mandible, 57
    pithecoid traits, 264
    place of origin, 267
    skull, 56, 269
    tympanic bullæ, 57, 60
Tarsius, alimentary canal, 270
    brain, 156, 158, 165, 270
    cæcum, 204
    claws, 135
    colon, 201
    dentition, 87
    digestive system, 199
    ear, 190

Tarsius, foot, 270
   limbs, 115, 121, 270
   liver, 207
   molars, 89
   nasal cavity, 177, 180
   ossicles, 192
   placentation, 217, 270
   reproductive system, 210
   skull, 56, 269
   special sense organs, 270
   sublingua, 195
   tongue, 194
   uterus, 217
Tarsus in monkeys, 130
   in Tupaiidæ, 230
Taxonomic values, 46
Teetee monkeys, 31
Teeth, evidence of, 69. *See* Dentition
Tegula, 104
Tertiary epoch, 20, 21
Testicles, descent of, 209
Theromorph reptiles, 19
Third trochanter, Hapalidæ, 128
Thumb, opposition of the, 126, 138, 228
Tongue, the, 192
   Anthropoidea, 195
   Catarrhine monkeys, 195
   Hemigalago, 193
   lemurs, 77
   in marmoset, 195
   in *Tarsius,* 194
   in tree-shrews, 245
Tree-shrews, 38
   affinities of the, 248
   relation of, to Primates, 222
Triconodonta, 19
Trigone, 72
Triplets in *Hapale,* 218
Tritubercular theory, 10
Tuberculum sextum, 77
Tupaia, colon, 201
   eye, 244

Tupaia, muscular system, 246
   ossicles of middle ear, 245
   placentation, 244
   prostate gland, 242
   retina, 244
Tupaiidæ, 223
   carpus, 227
   dentition, 231
   digestive system, 239
   limbs, 133
   tarsus, 230
Tupaioidea, 222
   affinities, 248
   brain, 234
   dentition, 231
   reproductive system, 242
   uterus masculinus, 243
Twilight vision, 182
Twins in *Hapale,* 218
Tympanic bulla, 46, 55
     in Lemuriformes, 46
     in Lorisiformes, 46
     in New World monkeys, 64
   chamber, 6
   ring, 40, 46

Uterus in Anthropoidea, 213
   in *Hapale,* 213
   in Lemuroidea, 212
   in *Tarsius,* 217
   masculinus in Tupaioidea, 243

Van Kampen, P. N., 46
Vermiform appendix, 204
   process of *Phascolomys,* 205
Vogt, O., 155

Werth, E., 95, 98
Winge, H., 47
Wislocki, G. B., 179, 209, 210
Woollard, H. H., 117, 182, 183, 184, 188, 191, 194
Wortmann, J. L., 74